MURDER & MAYHEM ON THE TEXAS RAILS

JEFF CAMPBELL AND THE
INTERURBAN RAILWAY MUSEUM

Published by The History Press
Charleston, SC
www.historypress.com

First published 2022

Manufactured in the United States

ISBN 9781467151450

Library of Congress Control Number: 2022933326

Notice: The information in this book is true and complete to the best of our knowledge. It is offered without guarantee on the part of the authors or The History Press. The authors and The History Press disclaim all liability in connection with the use of this book.

CONTENTS

ABOUT THE INTERURBAN RAILWAYS

Many of the stories in this book are related to the Texas Electric Railway, an interurban system that served North Texas from 1908 through 1948. (*Interurban* means "traveling between cities"; an interurban system was an electric street railway with more than half of its rail lines outside of city limits.) The Texas Electric Railway spanned 266 miles, making it the largest interurban railway system west of the Mississippi River.

The Texas Electric Railway was the result of the consolidation of the Denison and Sherman Railway, the Texas Traction Company, the Southern Traction Company and the Dallas Traction Company.

With a main hub in Dallas, the Texas Electric Railway stretched south to Waco, north to Denison, west to Fort Worth, southeast to Corsicana, east to Terrell, southwest to Cleburne and northwest to Denton. The Texas system was one of hundreds of interurban systems across the United States.

Interurban systems began springing up in the United States in the late nineteenth century. According to *American Heritage* magazine:

> *A United States congressman, Charles L. Henry of Indiana, coined the word* interurban *to describe the two-mile electric line he opened in the spring of 1892 between Anderson and North Anderson, Indiana, but the fifteen-mile East Side Railway, which began operation between Portland and Oregon City, Oregon, in February of 1893, is usually regarded as the first true interurban. Others soon appeared in almost every part of the United States, and by the turn of the century, the boom was on.*[1]

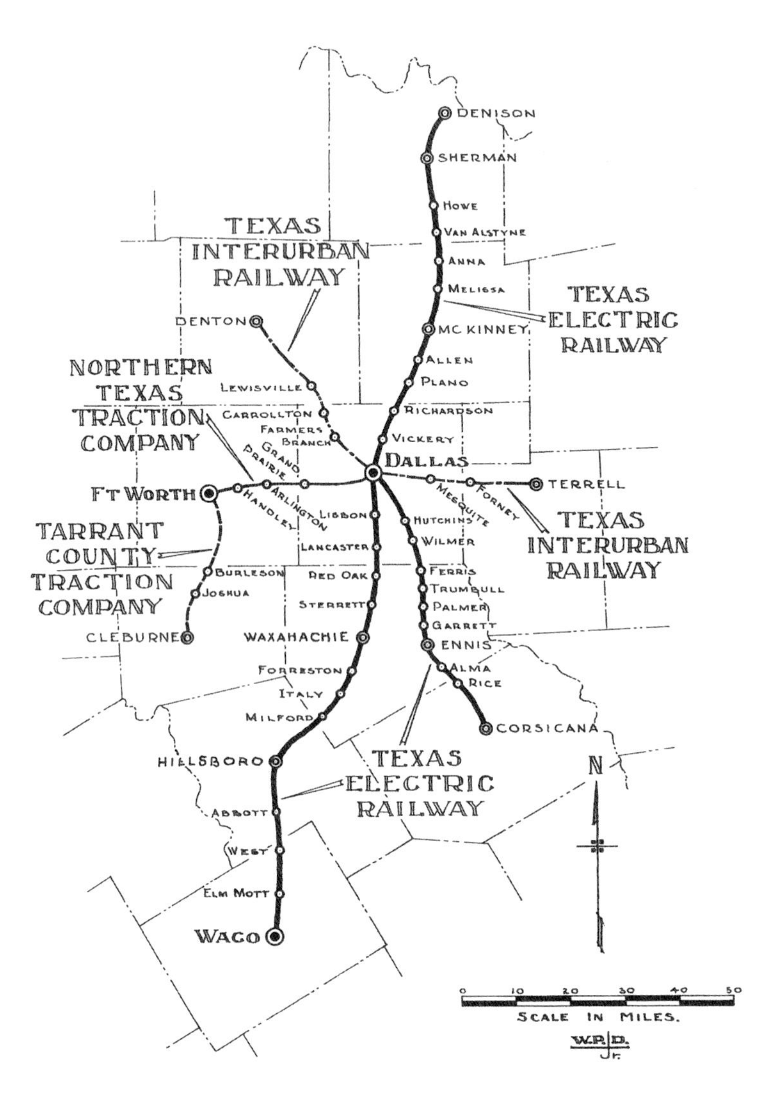

Map of the Texas Electric Railway System that ran from 1908 to 1948. *Courtesy of the Plano Conservancy for Historic Preservation.*

However, by the 1950s, most all of the interurbans were defunct, due to "R&R" (Rubber and Roads). Americans wanted automobiles and the freedom of the open road. They had little desire for the confines of mass transit.

Other stories contained in this volume come from across the state and include passenger trains, freight trains and trolleys, along with the interurbans. I hope you enjoy these stories of murder and mayhem on the Texas rails.

ACKNOWLEDGEMENTS

Jeff Campbell would like to thank:

Mary Jacobs for editing this book.

Hunter Herring for writing the story "Railway Employees: Into the Jaws of Death," researching "New Number Same Result" and writing "The 1914 Woodlake Interurban Wreck."

Austin Ng for researching and writing "The Mystery of John H. Dillehay's Death" and researching for "The Costs of Cost Cutting."

Debbie Calvin for her assistance in assembling the *Mayhem on the Interurban* exhibit for the Interurban Railway Museum and the photographic research for this book.

Druce Reiley Associates and the creative team at the Interurban Railway Museum for the museum exhibit *Mayhem on the Interurban.*

Cheryl Smith and the Haggard Library for their research assistance.

Harold Larson, president of the board of the Plano Conservancy for Historic Preservation, for his support of this project.

And Newspapers.com, an excellent resource for anyone conducting historical research.

INTRODUCTION

America has a long-romanticized relationship with trains. That relationship goes back to May 10, 1869, when that last golden spike was struck with a silver hammer at Promontory Summit, completing the first intercontinental railroad.

Since then, movies have been made, books have written and model railroad sets have been sold to wide-eyed children and adults. Songs have been written, like the "Wabash Cannonball," the "Chattanooga Choo-Choo," "City of New Orleans," "Folsom Prison Blues," "Rock Island Line" and thousands of others.

There was a time when passenger trains were the luxurious way to travel. The advertisements from the Santa Fe Railroad, in particular, tugged at the vacationer's heart.

Although passenger train travel has declined over the years, our collective romance continues into the new millennium. Tourists joyfully ride the streetcars in New Orleans and the cable cars in San Francisco. There are numerous successful train museums across the country, including the one this author manages, the Interurban Railway Museum in Plano, Texas.

Every day, the museum's staff sees the excitement and wonder in faces both young and old. The model train layout, depicting 1920s Plano, always draws a crowd. However, the highlight of any visit to the museum is a tour of historic car 360. Car 360 is constructed of wood and metal, built by the American Car Company in St. Louis in 1911. Similar to a trolley car, car 360 served the Texas Electric Railway until 1948.

Above: Southern Pacific's *The Owl*, train no. 17, northbound, headed by diesel locomotive no. 203, crossing the Trinity River Bridge en route to Dallas at sunrise on the morning of June 22, 1952. *Courtesy of the Museum of the American Railroad, Burt C. Blanton Collection, Portal to Texas History.*

Left: The Interurban Railway Museum in Historic Downtown Plano, Texas. *Photograph courtesy of the Plano Conservancy for Historic Preservation.*

As visitors step into the bright-red car, they are transported into the golden age of rail transportation, surrounded by resplendent, varnished wood walls, wrought-iron luggage racks and plush leather seats. It is a testament to craftsmanship and preservation. The environment heightens a visitor's love of trains.

However, there is a flip side to the nation's romance of trains—a dark side. There are horrific tales of train wrecks, robberies and murder. Mayhem reigned with this new technology, as trains wrecked and crashed into unsuspecting motorists. The railroad system gave those "up to no good" new opportunities to rob, steal and murder. Electric trains were even more dangerous; this new technology maimed and killed the uninstructed and the careless. The following tales will shine a light on the dark side of Texas railroad history.

This page: America has an ongoing romance with streetcars and trolleys, such as the San Francisco cable cars *(right)* and New Orleans streetcars *(above)*. Cable Car Days, *from the Johnnie J. Myers Archives, photograph by Jeff Campbell. New Orleans streetcar photograph courtesy of the Johnnie J. Myers Archives.*

Malcolm Len Morrow and his sister Martha Morrow Gamblin in 2015, discussing how their dad, motorman M.A. Morrow, worked for the Texas Electric Railway. *Photograph by Jeff Campbell.*

1
A FEW DEFINITIONS

The following are some helpful definitions of terms used in the interurban stories that may be unfamiliar to some readers.

- car: Short for *interurban car*. Interurban cars were essentially souped-up, sometimes lavishly appointed trolley cars. Most had powerful motors and high-speed gearing, and they were usually larger and heavier than their city cousins. Cars were numbered, like "car 360," which is on display at the Interurban Railway Museum in Plano.
- motorman: A rail vehicle operator.
- siding: A short section of railway track connected to a main line, usually parallel to a rail station, that allowed other cars trains on the same line to pass.
- trolley:
  - › A streetcar powered electrically through a trolley, also referred to as a trolley car.
  - › A device that carries an electric current from an overhead wire to an electrically driven vehicle.

2
MOUNTAINEER MADMAN

Samuel A. Cole lived a typical American life—until November 3, 1928, when his life was irrevocably shattered.

A motorman for the Texas Electric Railway, Cole married Ruby Bates in Ellis, Texas, in 1916. The couple had three children and lived a typical American life. Samuel's father even ran a grocery store.

On that fateful autumn day, Cole was shot and killed by Dewey Hunt in an attempted armed robbery on Cole's Dallas Interurban car.

> *Cole was found dying in his car, which had rolled to a stop on Lindsley Avenue at the intersection of Cameron* [Avenue]. *He had been shot twice, once through the left side and once in the head. He died while being taken to the hospital. The discovery was made at 8:30 o'clock by H. Compton, streetcar motorman on the same line, who investigated when he saw Cole's car stopped.*[2]

At the time of the robbery, Cole was protecting his nightly receipts, consisting of $8.00 in cash and $2.40 in streetcar tokens. Cole had once been a victim of hijackers and vowed to his coworkers he would never be robbed again.

On November 6, 1928, Hunt was formally charged with the murder of Samuel A. Cole. The murderous actions by Hunt left Ruby Cole a widow and her three children fatherless—because of a murderer's attempt to steal $10.40.

DEWEY HUNT OF THIS COUNTY KILLS TEXAN; BEING HELD

Authorities Will Probe Record of Escaped Asylum Patient.

"TOUGHEST" PRISONER IN DALLAS

"Too tough" to be tried, is Dewey Hunt, late of Tennessee, charged with murdering a street car conductor in Dallas, Tex. He battled officers for 30 minutes in the jail, was subdued with tear gas, brought into the courtroom shackled and strapped, only to raise such a commotion that it was impossible to proceed with the case. Above, Hunt, as he snarled at the judge.

Left: On November 6, 1928, Hunt was formally charged with the murder of Samuel A. Cole. *Courtesy of the* Jackson Sun *and Newspapers.com.*

Right: The criminally insane "Toughest Prisoner in Dallas." *Courtesy of the* Kingsport Times *and Newspapers.com.*

As the holidays approached and the temperatures dropped, Cole was laid to rest. Reverend Walter H. McKenzie gave the rites at Hillcrest Burial Park Cemetery in Waxahachie, Texas. Cole's pallbearers were A.S. Green, W.N. Vickery, W.J. Vickery, D.H. Cartwright, P.E. Moore and H.C. Griffin. Samuel's parents, John and Lucy Green, looked on, carrying the burden of parents who outlive a child.

As for Dewey Hunt, the folks from his native Tennessee had to know this was an inevitable conclusion. On August 19, 1928, the *Jackson* (TN) *Sun* newspaper reported on one of Hunt's earlier crimes. Hunt had hijacked a car with two passengers at gunpoint on August 13 and was arrested.

Hunt did not adjust well to his confinement. By Tuesday, he had become violent and began screaming incoherently. A doctor was called, but Hunt refused to let anyone in his cell. By Friday, he was tearing plumbing fixtures off the wall and hurling them at anyone who came near. Finally, the authorities used tear gas to subdue Hunt. He was handcuffed, given an opiate and shipped off to Western State Hospital, an asylum in Bolivar, Tennessee.

Before the calendar could flip from August to September, Hunt escaped from the asylum. Like many criminals before him, he headed west to Texas. Hunt found himself in San Antonio and, in another bizarre turn of events, joined the army. Hunt found the army barracks as confining as the asylum. He soon went AWOL, deserting the army, a crime he later confessed to. Next, he headed east to Houston and robbed a streetcar motorman. Finally, Hunt went north to Dallas, where he murdered Cole.

During his four-day trial in early 1929, Hunt battled with guards, throwing bottles of milk at deputies, screaming and cursing in the courtroom. Authorities gagged him and strapped him to a chair. Hunt's antics would not help him avoid prosecution, as they did in Tennessee.

A jury found him guilty of murder without malice. Hunt was sentenced to death on circumstantial evidence. There was a strand of blue yarn found on Cole's car that matched Hunt's bloody blue sweater found in his room. A Dallas chemist testified to the match of the fabric. There was also a trail of blood from the car to a house where Hunt had stopped to ask a woman to call a taxi for him. The woman, Miss Elreno Petty, was able to identify Hunt

1 PLACE OF DEATH
STATE OF TEXAS
COUNTY OF Walker

TEXAS STATE DEPARTMENT OF HEALTH
BUREAU OF VITAL STATISTICS
STANDARD CERTIFICATE OF DEATH

57890
Registrar's No. 1289

CITY OR PRECINCT NO. Huntsville No. Street Texas Prison System
If in an Institution, give name of Institution instead of Street and No.

Length of residence in city where death occurred ... yrs. ... mos. ... days. ? How long in U. S. if foreign born? ... yrs. ... mos. ... days.

2 FULL NAME OF DECEASED DEWEY R. HUNT.

Residence: No. ... Street ... Texas Prison.
If non-residence give city, or town and state

PERSONAL AND STATISTICAL PARTICULARS

3. SEX Male
4. COLOR OR RACE White
5. Single Married Widowed Divorced (Write the word) Single
5a. If married, widowed, or divorced HUSBAND of (or) WIFE of
6. DATE OF BIRTH (month, day, and year) 1906
7. AGE 27 Years Months Days If LESS than 1 day, ... hrs. or ... min.
OCCUPATION
8. Trade, profession, or particular kind of work done, as spinner, sawyer, bookkeeper, etc.
9. Industry or business in which work was done, as silk mill, saw mill, bank, etc.
10. Date deceased last worked at this occupation (month and year)
11. Total time (years) spent in this occupation
12. BIRTHPLACE (city or town) (State or country) Tenessee
FATHER
13. NAME L.B.Hunt
14. BIRTHPLACE (city or town) (State or country) Tennessee
MOTHER
15. MAIDEN NAME Matilda Melton
16. BIRTHPLACE (City or town) (State or county) Tennessee
17. INFORMANT Taken from Criminal Records
(Address) Insitution
18. BURIAL, CREMATION, OR REMOVAL Place Routine Date ... 19
19. UNDERTAKER none J. G. Ashford Jr
(Address)
20. FILE DATE AND SIGNATURE OF REGISTRAR 2/31/33 19 [signature]

MEDICAL CERTIFICATE OF DEATH

21. DATE OF DEATH (month, day, and year) 12-29-33, 19
22. I HEREBY CERTIFY, That I attended deceased from 12-29-33, 19 ... to 12-29-33, 19
I last saw h im alive on 12-29-33, 19 ...; death is said to have occurred on the date stated above, at 12-05 a. m.
The principal cause of death and related causes of importance were as follows: Date of onset
Executed by order of court.
Other contributory causes of importance:
Name of operation ... date of ...
What test confirmed diagnosis? ... Was there an autopsy? ...
23. If death was due to external causes (violence) fill in also the following:
Accident, suicide, or homicide? ...
Date of injury ..., 19 ...
Where did injury occur? ... (Specify city or town, county, and State)
Specify whether injury occurred in industry, in home, or in public place.
Manner of injury ...
Nature of injury ...
24. Was disease or injury in any way related to occupation of deceased?
If so, specify ...
(Signed) Lynn Hellum M. D.
(Address) ...

Dewey Hunt's certificate of death from the Texas State Department of Health. *Courtesy of the Plano Conservancy for Historic Preservation.*

MAN DIES WITH SMILE ON LIPS

Dewey Hunt Pays Extreme Penalty for Slaying Of Motorman

With an unsettling smile on his face and a cigar in his mouth, Hunt strutted into the chamber of death. *Courtesy of the* Brownsville Herald *and Newspapers.com.*

as the man who had approached her. The taxi driver also identified Hunt.

In October 1931, Hunt went on trial for a second time. The court of criminal appeals stated that a murder such as this could not be punishable by death and should only have a sentence of a maximum five years.

Hunt was convicted of murder a second time. The verdict was appealed to the United States Supreme Court. The court declined to assume jurisdiction with the case, and the verdict stood. Hunt would spend five years on death row at the penitentiary in Huntsville, Texas.

On December 29, Hunt was executed for the death of Samuel A. Cole. The *Corsican Daily Sun* reported, "Dewey Hunt's Five-Year Fight Is Ended with Jaunty Walk to Chair."[3] The *Knoxville Journal* headline read: "Mountaineer Goes to Chair."[4]

With an unsettling smile on his face and a cigar in his mouth, Hunt strutted into the chamber of death. At Hunt's request, a fellow death row inmate, Paul Mitchell, played a mountain breakdown called the "Chicken Reel" on harmonica. Hunt was electrocuted and met his death with that same unsettling smile on his face. Dewey Hunt was buried in Peckerwood Hill Cemetery (now known as Captain Joe Byrd Cemetery), the prison cemetery at the Huntsville Penitentiary.

3
CAPTAIN JOE BYRD CEMETERY

Why was Dewey Hunt buried in the Captain Joe Byrd Cemetery and not in his native Tennessee? Hunt's only next of kin was his mother, and he had stated, "I wanted to spare her the ordeal. She won't know until after I am electrocuted."[5]

When the families of the deceased can't afford a burial or do not claim the body, the prisoner's remains end up in the Captain Joe Byrd Cemetery. The cemetery was originally referred to as "Peckerwood Hill." *Peckerwood* was an insult aimed at poor Whites by Black Americans. The twenty-two-acre site was donated to the state (by Sanford Gibbs and George W. Grant) in the 1850s, only because prison officials already had—by mistake—used it for burials.[6]

Much later, the cemetery was named after Captain Joe Byrd, "an assistant warden at the Walls Unit who, in the 1960s, initiated a cleanup of the neglected grounds."[7] Captain Joe also flipped the switch of "Old Sparky," the oak electric chair that electrocuted 361 people from 1924 to 1964.[8] As morbid as this all sounds, Captain Joe was known for the respect and dignity he afforded death row inmates and their families.

It's hard to establish how many prisoners are buried at the cemetery. Early markers were made of wood and rotted away over time. During the 1980s and '90s, the concrete crosses included only prison numbers and dates of death. If the inmate was executed, the headstone bore the letters "X" or "EX," or a prison number beginning "999"—the designation for death row.[9]

A 1995 photograph of Captain Joe Byrd Cemetery. *Courtesy of the* Abilene Reporter *and Newspapers.com.*

In 2012, prison officials verified the bodies of 2,100 inmates who were buried at the cemetery, but they say there may be additional graves. Professor Franklin T. Wilson, an assistant professor of criminology at Indiana State University, photographed every headstone and estimated that there were more than 3,000 graves.[10] Each year, approximately 100 inmates are buried in the cemetery.

Although not all the prisoners buried at Captain Joe Byrd Cemetery were the "worst of the worst," they were all poor, alone or both.

4

REVENGE OF THE MOTORMAN

Harry C. Turnpaugh had a good job as a motorman for the North Texas Traction Company. His route ran between the cities of Fort Worth and Dallas. On February 26, 1921, Motorman Turnpaugh shot and killed Ernest T. Noel near the Tarrant County Courthouse. It was a Saturday morning on Main Street in Fort Worth. What was Turnpaugh's motivation?

Turn the calendar back to February 9 of the same year. Noel divorced his wife on that day. Mrs. Noel was granted custody of their two young children. Two weeks later, on February 23, Noel shot Mrs. Noel in the back at her home on 804 South Main Street. Mrs. Noel's maiden name was Turnpaugh—Inez Turnpaugh, Motorman Turnpaugh's sister.

On February 24, the *Fort Worth Star Telegram* reported that Turnpaugh told the newspaper:

> *The shooting was the result of Mrs. Noel not agreeing to return to or remarry Noel and was not caused by a "third party." Noel had called at the house to insist on Mrs. Noel to return to him and that he shot her when she refused, there was no other man in the case as was charged by Noel. The trouble had been going on for about 18 months, and he had repeatedly threatened her.*[11]

Ernest Noel had a different story:

> *I wanted her to remarry me because of our boys; recently, I went out of town and thought that she had agreed not to get the divorce. We had some*

> *trouble over a boarder who was responsible for separating us. She liked him and wouldn't give him up. The first time I had seen her since the divorce had been granted was Wednesday night. I went to her house to see a man there who had nothing to do with our trouble. I did not go there to see my wife, but she came to the door and said the man did not want to talk to me. She started to discuss our trouble. I asked her to do one of two things. To remarry me for the sake of the children or to marry the other man. But she wouldn't do it. She said she didn't have to marry me or him either. She said that it wasn't any of my business what she did with the children.*[12]

Following the discussion, Noel shot Inez.

Inez was taken to All Saints Hospital, where she began her slow recovery from the gunshot wound. Ernest Noel was charged with assault with intent to murder and was released from the Tarrant County Jail on a $2,500 bond. Ernest must have felt relieved and free to be out of jail. However, the move made him a sitting duck for Motorman Turnpaugh.

Friday must have given Motorman Turnpaugh plenty of time to stew over the situation as he made the thirty-five-mile trip between Dallas and Fort Worth. By Saturday morning, he was ready to avenge his sister's shooting.

On the winter Saturday morning, Ernest Noel was driving in downtown Fort Worth, near his attorney's office. Turnpaugh was standing on the sidewalk as Noel turned in front of the Exchange State Bank.

Suddenly, Turnpaugh leaped onto the running board of Noel's car and quickly fired five shots from his pistol. Noel slumped over the steering wheel as the car came to a crashing halt. Noel was dead from three bullets to his head and neck. Turnpaugh, uninjured, calmly handed his pistol to a bystander.

Dan Dennard, a deputy constable, happened to be close by and escorted Turnpaugh to the district attorney's office. At the DA's office, Turnpaugh did not sign a confession but did admit that he killed Noel due to the cruel treatment of his sister. Turnpaugh was released on a $10,000 bond. A grand jury investigation was scheduled to begin the following Wednesday, March 2.

On March 19, the headline in the *Fort Worth Star Telegram* read "No Bill Is Voted for Noel Killing."[13] Assistant District Attorney W.R. Parker told the paper that the grand jury was unanimous in deciding that Turnpaugh should not be indicted.

Some would say that Motorman Turnpaugh got revenge. Others might say he got away with murder.

5
THE DEATH OF LARKIN GEORGE

The Houston and Texas Central Railway linking McKinney and Dallas arrived in 1872, and life changed dramatically in Plano. By 1873, the Texas Central Railway stretched all the way to the Red River, where it linked with the Missouri, Kansas and Texas Railroad, thus creating a railroad route connecting Houston, Texas, and St. Louis, Missouri.

Downtown Plano became the prosperous center of a booming farm economy, thanks to the emergence of cotton as a cash crop, cheap fencing in the form of barbed wire and the development of mechanized farm equipment. Visitors can still see pulleys and skylights in some of the buildings downtown, where cotton bales were lifted into the building for inspection and trade.

In 1908, downtown Plano got a second railroad stop: the Texas Electric Railway, a service that linked Denison and Dallas by way of Plano. The railway brought commerce to Plano, but it also brought crime and mayhem. There were several gruesome accidents involving the train, including a horrific story of a family in the neighborhood who went out for a Sunday drive. Their car was struck by a train, and two of the family members were killed instantly.

If you are a fan of *Downton Abbey*, you probably remember the scene in which the Dowager Countess (Maggie Smith) refuses to turn on an electric light. In the early years, people were frightened by electricity; in fact, it was dangerous because they didn't understand it. Maybe that's what led to the tragic end in 1909 of Larkin George, a young man in the employ of Phillpott Hardware Company.

Looking up in the tower room of the Texas Electric Railway Plano Station (now the Interurban Railway Museum). This is where Larkin George fell to his death. *Photograph by Jeff Campbell.*

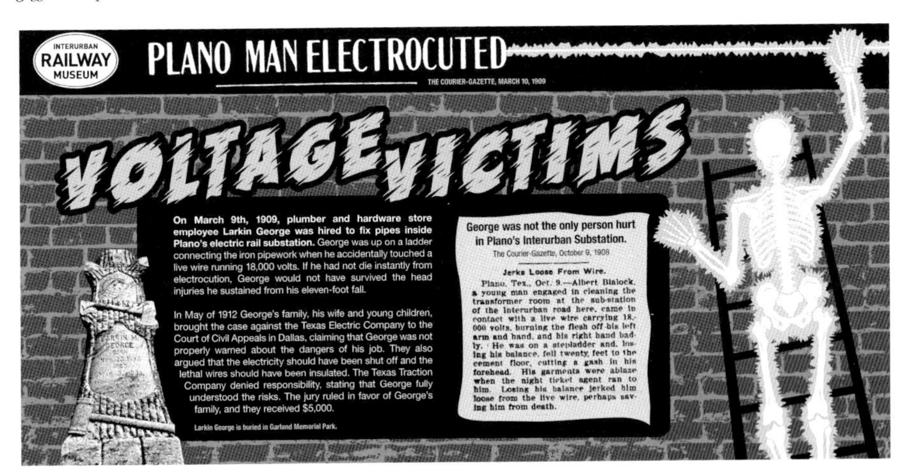

In May 1912, George's family, his wife and young children, brought the case against the Texas Electric Company to the court of civil appeals in Dallas, claiming that George was not properly warned of the dangers of his job. They also argued that the electricity should have been shut off and that the lethal wires should have been insulated. The Texas Traction Company denied responsibility, stating that George fully understood the risks. The jury ruled in favor of George's family, and they received $5,000. *Courtesy of the museum panel at the Interurban Railway Museum.*

It was about 8:30 p.m. when he was called in to work on connecting a pipe. The station had only been there for a year, and George did not possess very much electrical knowledge.

He accidentally touched a high-tension wire in the ceiling. George was zapped and hit the floor. Physicians called to the scene tried to resuscitate him, but George was dead.

He should have known to be more careful. Another young worker named Blalock was burned in the same location, just months before George's death. Blalock was cleaning the substation when he came into contact with a live wire carrying eighteen thousand volts. It burned the flesh right off his right hand and arm. He was on stepladder, the same as George, and fell twenty feet to the cement floor. He cut a big gash in his head, and his clothes were on fire. Luckily, the ticket agent was on duty; he came and pulled him away from the wire and put the fire out. George felt bad for Blalock but probably never thought he would be next. Larkin George left behind a wife and two children.

6

RAILWAY EMPLOYEES

INTO THE JAWS OF DEATH

The year is 1929. It's 7:50 a.m. on a pleasant September morning. A twenty-two-year-old man named Sanders Dowty, an employee of the Texas Electric Railway, is being crushed alive. A few moments ago, he had been enjoying a small breeze at the back of an express car and listening to the click of the wheels as the vehicle slowly lumbered down the railroad tracks of Waco. Now, he's trapped, pinned between the cool metal of the car and the rough brick wall of the interurban station. As Dowty begins to lose consciousness, a horrible crunching sound reaches his ears, and he loses feeling in the lower half of his body. Later, at Colgin's Sanitarium, doctors worked quickly to save his life, taking X-rays of his crushed pelvis and various internal injuries before performing an emergency operation.

Sanders Dowty was just one of many seriously injured railway workers in the early twentieth century. Railway employees worked long, grueling shifts—often thirteen, fourteen and sixteen hours a day, depending on the job—for a measly twenty-one to twenty-four cents an hour. Exhausted workers pushed themselves to handle and navigate around heavy machinery and high-voltage electricity, operating under the Texas Electric Railway's strict schedule of arrivals and departures. Work conditions weren't always ideal. If an accident occurred and someone was seriously injured, the trains were often between cities, miles away from the nearest doctor.

As the main mode of transportation in the nineteenth century, trains enabled people to travel long distances. They were faster than a horse and buggy and cheaper than a personal automobile. Trains also spurred

Railway work was extremely dangerous—even more so at night. Pictured is a Texas Electric Railway nighttime switching operation at Oak Cliff Junction near Dallas, Texas. *Photograph courtesy of the Johnnie J. Myers Archives.*

economic growth by providing jobs to civilians and bulk transportation of food, goods and building materials. At the same time, trains were fraught with danger for the employees who worked with them. As rail use and track mileage expanded across the United States, so, too, did railway worker injuries and fatalities. In 1900, 1 in every 28 railroad employees were injured on the job; 1 in 399 died. Eight years previously, Massachusetts congressman Henry Cabot Lodge mournfully compared the rate of injuries and deaths of railway workers to the carnage of war: "The object of trainmen is to carry on safely the railway traffic of a great country. Yet they suffer as if they were fighting a war, and the percentage of loss to numbers employed, if not so high as with soldiers, is frightful enough."[14]

Statistically, it wasn't uncommon for brakemen to lose fingers or an entire hand when coupling trains if the cars rolled too quickly and slammed together or if their bumpers misaligned. Legs and feet were often crushed or entirely torn from the body under the heft of rolling train wheels, whether the car was moving at a snail's pace or speeding by. Skulls were cracked in falls from the tops of moving vehicles, probably in emergency dismounts. Some employees even died in bad collisions or derailments. Cuts, bruises, burns, electric shock—these were the war wounds of railway workers on their battlegrounds of rails and ties.

In addition to vehicle-related accidents on the railway, workers were also prone to the natural hazards of human existence, such as heart attacks, strokes or seizures, sometimes miles deep in the countryside with no city or hospital in sight. In the early years of the Industrial Revolution, scenes like this were commonplace:

> [The worker is] *usually tied up with rope, old rags, soiled handkerchiefs, or anything else lying about, lifted into the first train, possibly sometime after being hurt, with his crushed members dangling behind him, unsupported; then sent along the road many miles in a cold damp car, each start and jar of which would almost close the scene…the crushed arm or leg so mixed up with clothing, gravel, sticks, etc., that the whole mass looks like nothing but bloody rubbish.…He has been jostled and bled to death, and so he dies.*[15]

These long stretches of rural track were so dangerous that at the start of the Civil War, large steam railway companies would frequently establish new medical facilities, complete with contracted railway physicians and surgeons, along the lines. These medical employees were so essential to operations that they were considered railway employees themselves.

When a passenger or worker was injured or maimed, help was never far away, which helped lessen legal liabilities for the railway company and prevented empty shifts the day after an accident. Some companies even established mutual benefit associations for workers that covered the treatment of injuries long before healthcare was provided by most industries. Unfortunately, employees who worked for the Texas Electric Railway were not so lucky.

Company Rules and Regulations

The Texas Electric Railway Rule Book was the company's central approach to safety. As stated in the first few pages of the booklet, "Observance of the rules is essential to the safety of passengers and employees and to the protection of [company] property.…Employee, in accepting employment, assume its risks." In general, the book provided basic instructions necessary for employees, from motorman to dispatcher, to fulfill their duties on a daily basis.

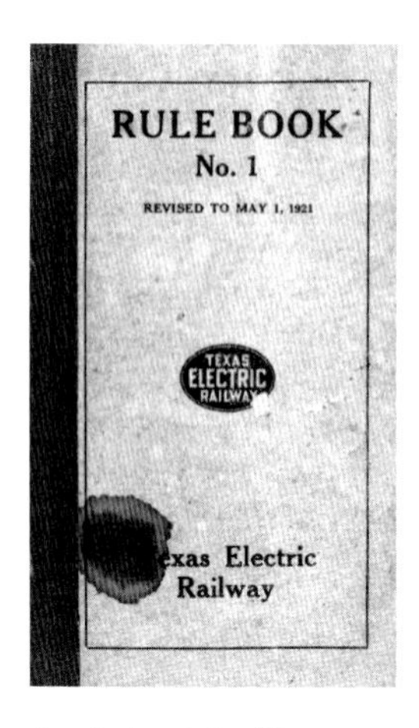

A rule book for Texas Electric Railway Employees. *Photograph courtesy of the Johnnie J. Myers Archives.*

The importance of following the rules was stressed to the point that employees were required to carry a copy of this book at all times while on duty. It was expected that all employees of the Texas Electric Railway knew the company rules. In addition, they were required to keep an eye out for any special instructions that were given to them by the head of their department on a daily basis.

Aside from this, safety is mentioned briefly in two half-page sections on pages 66 and 67 of the rulebook, under the subheading "Accidents and Personal Injury." In the instance of an accident, no matter how minor, employees were instructed to "render all assistance necessary and practicable," never leaving an injured person without care. Minor scrapes could be treated with medical kits that were kept in each substation. In the instance that an employee was burned, they were either transported to the nearest substation or someone was sent for the kit to be brought to the site of the accident.

Conductors and motormen were to make an immediate report to the dispatcher if the injury was at all serious. If dispatch could not be reached for further instruction, the person could be carried to the nearest town for medical assistance. Once a physician arrived on the scene, railway employees could not authorize any medical treatment on behalf of the injured, except in extreme circumstances, nor could they pay them a visit in the hospital afterward without express permission from the company's claim department. A full written report, with names and addresses of all involved parties, including witnesses, license plate numbers, badge numbers and so on, were required to be filed as soon as possible. Afterward, employees could not speak to anyone regarding accidents unless they were an authorized representative of the Texas Electric Railway Company.

At the back of the rule book, more immediate instructions were available for accidents that required immediate action. Pages 85 to 90 can be referenced for "Resuscitation from Electrical Shock"; page 91 for "Emergency Treatment for Burns" and "Burns from Fire, Electric Arc, Explosion of Gases, or Scald"; and page 92 for "Burns from Caustic, Lye, or Strong Ammonia," "Burn by Acid" and "Treatment of Eyes for Burn by Electric Flash."

PHYSICAL AND FINANCIAL DAMAGES

That much danger to life and limb is incurred by those who are thus engaged in this special vocation is self-evident, and even in localities where the strictest supervision and management are employed to avoid danger, the number of accidents is painfully large.
—Clinton Bradford Herrick, MD[16]

The most frequently injured railway employees were brakemen, followed by switchmen, firemen and engineers, due to the hazards that came with each of their jobs. Injuries were typically received during car coupling, when jumping on or off moving trains, falling off cars and accidents that involved crashes and derailment. (Passengers were more often hurt in derailments or collisions, rather than falling from trains like railway employees.)

Workers hurt on the job (or the widows and heirs of workers) could choose to sue the railway company for damages, but winning wasn't easy. The company could simply point back to the rule book and the signature of the employee on the front page, where they agreed that in accepting employment, they assumed the risks that came along with it. Long schedules and hazardous work were often causes of contempt among Texas Electric Railway employees. Workers organized many strikes and walk-outs throughout the company's history. Any liability on behalf of the company for potentially unsafe working conditions was dodged, and blame was usually placed squarely on the shoulders of that particular worker or any other employees who were present. In 1900, a little less than half of the families of fatally injured railway workers received any financial compensation at all. The families who were awarded usually received about half a year's pay.

Between 1911 and 1921, forty-four states passed workmen's compensation laws that automatically granted money to the victim or their family at a higher fixed rate. (To this day, Texas is the only state that allows employers to choose whether or not they want to provide compensation.) Large companies were presented with a choice: care about the safety of employees at work or pay up. Suddenly, employers began guarding sources of high voltage, requiring that workers protect themselves with safety equipment and encouraging managers to actively search for and report hidden dangers. As a result, railway work accidents, injuries and fatalities declined steadily after 1910.

In January 1931, in the middle of the Great Depression, the Texas Electric Railway Company went into receivership, as it teetered on the edge of bankruptcy. In a controversial move to save some cash, the company decided that rather than continuing to utilize the typical motorman and conductor duo, passenger cars would switch to one-man operations. The conductor position was eliminated, and motormen were simply referred to as operators. Most rear exits were shuttered (the only exceptions being ex–Southern Traction and ex–North Texas Traction cars), and the entrance was moved to the front right side of the train, next to the operator's seat. Many employees did not like this change and considered it a safety hazard for themselves and their passengers. The company altered the front side windows, ensuring that they were large enough to prevent any blind spots. Indeed, across the United States, regardless of any adjustments companies made to lessen the dangers of one-man operations, workers would band together in unions to resist and strike.

In the following years, crashes continued to occur, seemingly increasing in frequency and in the number of injuries. Three of the worst wrecks in the history of the Texas Electric Railway occurred in virtually the same spot in Denison in 1945 and 1946. The third hospitalized over forty people, many with serious head injuries, and prompted concerned legislators to ask that the company be investigated by the State Railroad Commission for inadequate safety precautions. In the end, the company's dedication to business over safety fanned the flames of its poor financial situation, contributing to the final end and abandonment of the railway in 1948.

Today, railroad work is still one of the most dangerous jobs in the United States. Even with modern technology and increased safety awareness, employees still receive injuries while at work through no fault of their own. Unions warn workers to only visit a company doctor once, file injury reports quickly and thoroughly, collect photographic evidence, gather witness contact information and even lawyer up, fearing unfair treatment from their employers. Railroad employees in Texas still do not fall under typical worker's compensation laws. Still, whether it is their love for trains or their dedication to public service and the benefits of railroads in society, workers push on, day in and day out.

Sources

Clinton Bradford Herrick, MD. *Railway Surgery: A Handbook on the Management of Injuries*. N.p.: William Wood and Company, 1899.

Economic History Association. "History of Workplace Safety in the United States, 1880–1970." https://eh.net/encyclopedia/history-of-workplace-safety-in-the-united-states-1880-1970.
Texas Department of Insurance. "History of Workers' Compensation in Texas." March 11, 2020. https://www.tdi.texas.gov/wc/dwc/history.html.

Thanks to Hunter Herring for researching and writing this story.

7
TEXAS ELECTRIC RAILWAY RULE BOOK NO. 1

For anyone employed by an interurban or streetcar company, electricity was a constant, treacherous risk. The Texas Electric Railway published *Rule Book No. 1*, for its employees, which devoted a section to electrical safety. The following section is taken from the 1921 edition.

RESUSCITATION FROM ELECTRICAL SHOCK
BY THE PHONE PRESSURE METHOD, RECOMMENDED
BY NATIONAL ELECTRIC LIGHT ASSOCIATION.

FOLLOW THESE INSTRUCTIONS EVEN IF VICTIM APPEARS DEAD.

I. FREE THE VICTIM FROM THE CIRCUIT IMMEDIATELY.

Quickly release victim from current, being careful to avoid receiving a shock. Using any dry non-conductor (rubber gloves, clothing, rope, board) to move either the victim or the conductor. Beware of using metal or any moist material. If necessary, shut off the current.

II. INSTANTLY ATTEND TO THE VICTIM'S BREATHING.

1. As soon as the victim is clear of the conductor, rapidly feel with your finger in his mouth and throat and remove any foreign body (tobacco, false

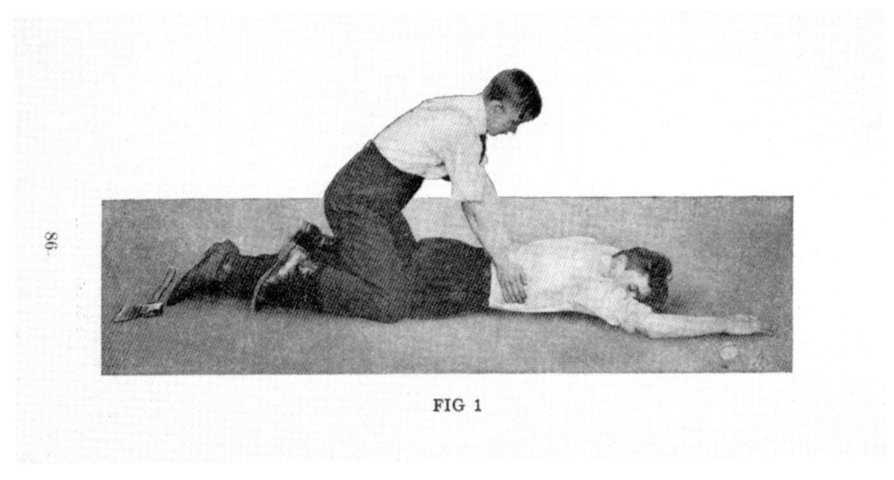

First aid directions from the "Rule Book for Texas Electric Railway Employees." *Photograph courtesy of the Johnnie J. Myers Archives.*

teeth, etc.). If just touching the lowest rib, the thumb alongside of your fingers; tips of fingers just out of your sight, as in fig. 1.

a) While counting one, two, and with arms held straight, swing forward slowly so that the weight of your body is gradually, but not violently, brought to bear on the patient. (See fig. 2). This act should take two or three seconds.

b) While counting three, immediately swing backward so as to remove the pressure, thus returning to the position shown in fig. 3.

c) While counting four, five—rest.

d) Repeat deliberately twelve to fifteen times a minute the swinging forward and backward—a complete respiration in four or five seconds. Time with your breathing.

e) While counting four, five—rest.

f) Repeat deliberately twelve to fifteen times a minute the swinging forward and backward—a complete respiration in four or five seconds. Time with your breathing.

g) As soon as the artificial respiration has been started, and while it is being continued, an assistant should loosen any tight clothing about the patient's neck, chest or waist. Keep patient warm.

2. Continue resuscitation (if necessary, four hours or longer) without interruption until natural breathing is restored or until a physician declares rigor mortis (stiffening of the body) has set in. If natural breathing stops after being restored, use resuscitation again.

3. Do not give any liquid by mouth until the patient is fully conscious. Place ammonia near the nose, determining safe distance by first trying how near it may be held to your own. Assistant should hit patient's shoe heels about twenty (20) times with a stick or something similar and repeat the operation every five minutes until breathing commences.
4. Give the patient fresh air, but keep him warm. When patient revives, keep him lying down and do not raise him. If doctor has not arrived, give patient one teaspoon of aromatic spirits of ammonia in a small glass of water if he can swallow.
5. Carry on resuscitation at closest possible point to the accident. Do not move patient until he is breathing normally without assistance. If absolutely necessary to move, he should be placed on a hard surface, such as a door or floor of conveyance. Do not stop or interrupt resuscitation for an instant.

III. SEND FOR DOCTOR.
If alone with victim, do not neglect to call a doctor; start at once, the first few minutes are delay.
The prone pressure method of artificial respiration described in rules (section II) is equally applicable to resuscitation form electrical shock as well as all cases of suspended respiration due to drowning, inhalation of gas, smoke or fumes or to other causes.

DO NOT INTERRUPT RESUSCITIATION UNTIL PATIENT BREATHES OR RIGOR MORTIS (STIFFENING OF THE BODY) SETS IN.

EMERGENCY TREATMENT FOR BURNS

Note: a complete emergency medical kit is carried in each substation. If possible, either take the patient to nearest substation or send for a kit. Use contents of kit in accordance with directions on cover of cabinet.
Burns are generally classed according to degree:

Simple reddening of the skin: 1st degree.
Formation blisters: 2nd degree.
Charring of the skin, up to complete destruction of the part: 3rd degree.

Burns of the 2nd and 3rd degree require immediate medical attention, especially if the area is large. A burn of the 1st degree may be very serious if sufficient area is affected. In severe burns, there is a liability of shock and prostration.

Burned portions should be excluded from the air, as contact with the air greatly increases the accompanying pain. Dry, charred skin is also very irritating to the burn. For this reason, an oil dressing is usually most satisfactory for emergency treatment, as it excludes the air, at the same time, softening the burned tissues.

BURNS FROM FIRE, ELECTRIC ARC, EXPOSION OF GASES, OR SCALD.

Remove clothing from around the burn by cutting away with knife or scissors. If the clothing sticks, do not pull it off; cut around it, leaving it for the surgeon to remove.

For emergency treatment, a vegetable oil, such as raw linseed, olive oil, sweet oil or cotton seed oil, is excellent. Vaseline or petro-latum are very good. Mineral oil, such as lubricating oil, should only be used as a last resort.

Cover the burned portion with gauze; saturate with oil; cover this with a layer of absorbent cotton and bandage lightly.

If the burn is of a serious nature, obtain medical attention as soon as possible.

TREATMENT OF EYES FOR BURN BY ELECTRIC FLASH

Where the eyes have been exposed to electric flash or strong electric arc, bathe with witch hazel, allowing the liquid to flow freely through the eyes. Bandage sufficiently to exclude light. If no witch hazel is to be had, saturate the bandages with sweet oil.

Thanks to the Johnnie J. Myers Archives and Research Center at the Interurban Railway Museum for this content.

8
TRAINS, PAINS AND AUTOMOBILES

The electric railway system's launch in Texas in the early 1900s presented advantages and challenges for both passengers and transportation companies. Interurban lines ran through small towns, providing convenient travel, transit for goods and access to new electric technology. However, the electric railway's novelty led to misunderstandings and a lack of respect for its dangers. Residents ignored safety warnings about electricity and the railway, resulting in many accidents.

Accidents and collisions were the most frequent forms of mayhem on interurban lines. Today, we take for granted the gates and arms at railroad crossings. However, there was a time when they did not exist. According to railroad historian Dr. J. Myers, "Texas Electric Railway never installed automatic crossing gates or block signals. Too cost prohibitive."[17]

Automobile drivers raced across the tracks, neglecting to look out for interurban railway crossings or ignoring them altogether. Children and adults would sit or play on the tracks, unconcerned or unaware of the perils. Occasionally, miscommunication between motormen and dispatchers led to deadly head-on collisions. As with any new transportation technology, the benefits came with unpredictable consequences.

Automobile collisions with interurban cars were common in the early years of the electric railway system in Texas. Drivers often ignored warnings, attempting to cross the tracks before the train passed through. On October 6, 1912, Barney Cornelius and his family were driving home from Fort Worth when they collided with an interurban train car about a mile west

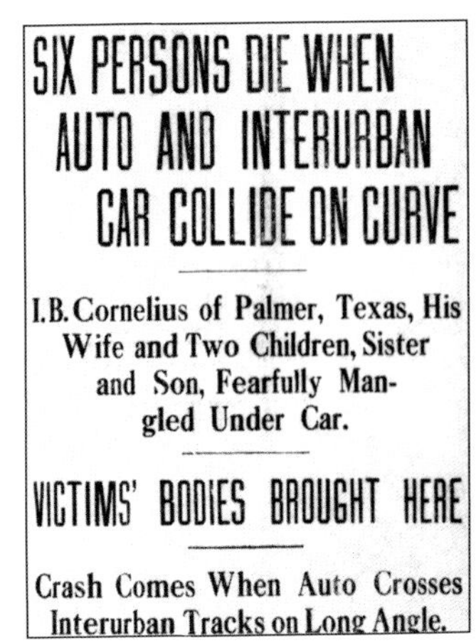

SIX PERSONS DIE WHEN AUTO AND INTERURBAN CAR COLLIDE ON CURVE

I.B. Cornelius of Palmer, Texas, His Wife and Two Children, Sister and Son, Fearfully Mangled Under Car.

VICTIMS' BODIES BROUGHT HERE

Crash Comes When Auto Crosses Interurban Tracks on Long Angle.

Left: A wrecked automobile on the Corsicana Line at Rice, Texas. Any time an accident occurred involving Texas Electric Railway cars, the staff photographer was required to document the wreckage, which was what occurred here. *Photograph courtesy of the Johnnie J. Myers Archives.*

Right: Headline in the *Fort Worth Record and Register*, October 7, 1912. *Photograph courtesy of Newspapers.com.*

of Arlington. Cornelius did not heed the oncoming train, which hit the car and sent it flying thirty feet. Cornelius, his wife, sister, two daughters and nephew all died instantly, according to a gruesome report from the *Wichita Daily Times*, "Captain Elliot, who lives near, hurried to the scene but found the victims beyond the need of human aid."[18]

On December 4, 1921, Manley Shanahan was out for a drive with his friends. The five teenagers were driving around Lisbon, about six miles south of Dallas. The car Manley was driving attempted to cross the tracks near the Lisbon railway station. The car and the five teenagers never made it to the other side of the tracks.

The car was struck by a southbound Texas Electric Railway Interurban car heading to Waco.[19] In an instant, sixteen-year-old Barney Cranford and fourteen-year-old Edwin Easter were dead. Two of the other young passengers suffered injuries.

Manley Shanahan, the driver, claimed he could not hear the approaching railcar because the automobile's side curtains were down. Older cars were "open," without windows on the sides. Canvas-like curtains were rolled down for protection from the cold and rain.

After the crash, Shanahan said, "I wished I had been killed, because people will say I was careless and was to blame for the whole business. The

Two Boys Killed In Interurban Smash

Southbound Interurban Car Hits Automobile at Crossing Out of Dallas, Killing Two and Injuring Others

Headline in the *Waco News Tribune*, December 5, 1921. *Photograph courtesy of Newspapers.com.*

folks here have not told me the condition of the other boys, but I could hear before they took me away from the wreck."

Shanahan walked away from the carnage with only a few scratches.

On a beautiful spring evening, May 2, 1924, three people were killed when an interurban car crashed into the Ford coupe of Perry Thomas. Thomas was traveling with two Corsicana public school teachers, Kathleen McKnight and Geddes Manning. They had left Corsicana at 4:30 p.m. and were headed to Dallas. Their trip came to a deadly end in Palmer, Texas, where they were struck by an interurban car. All three were killed instantly, their bodies mangled and the Ford coupe crushed.

On December 27, 1925, seven-year-old Catherine Langford was killed when a Fort Worth–Cleburne interurban car struck the automobile of Mr. and Mrs. Causey. Catherine's sister, eight-year-old Juanita, suffered a broken shoulder. The Causeys suffered only minor injuries. The Langford girls were returning to Cleburne after Christmas. The parents of the two girls were waiting in Burleson for an automobile that never arrived.

The autumn sun was just starting to rise on the east side of Waco as G.W. James drove to work on Monday, October 19, 1925. Perhaps Mr. James's mind was elsewhere, either reliving the weekend or planning the week ahead. Whatever the reason, he did not realize an interurban car was racing toward him as he crossed the track. The collision transformed Mr. James's truck into a fireball and propelled the vehicle two hundred feet down the track. A group of men raced to the truck to try to save him, but the searing flames kept them at bay. After the flames died down, they were able to pull Mr. James's charred remains from the wreckage.

EXTRA **Corsicana Daily Sun** EXTRA

THE ASSOCIATED PRESS PAPER

VOL. XXVI. NO. 83 TEN PAGES CORSICANA, TEXAS, FRIDAY, MAY 2, 1924 TEN PAGES PRICE FIVE CENTS

2 CORSICANA TEACHERS KILLED

OIL OPERATOR ALSO DEAD--INTERURBAN HIT AUTO NEAR PALMER

CRASH OCCURRED LATE THIS AFTERNOON

Dallas Tots, Aunt Killed on Way To Visit With Santa

Car-Interurban Collision Brings Death to Small Boys and Woman; Another Child Injured

Top: Headline in the *Corsicana Daily Sun*, May 2, 1924. *Photograph courtesy of Newspapers.com.*

Bottom: Headline in the *Waco News Tribune*, December 21, 1940. *Photograph courtesy of Newspapers.com.*

From 1911 to 1936, an interurban railway ran between Houston and Galveston. On August 23, 1934, a firetruck attempted to cross the track that stretched for fifty miles between the two cities. The driver of the firetruck saw the approaching interurban car and turned the wheel in an attempt to avoid a collision. Instead, the vehicle sideswiped the interurban car. The car derailed and came to rest in the center of the street. Three firemen were injured, and a fourth, E.J. Rogers, was hurled forty feet to his death.

December 20, 1940, brought another Christmastime tragedy. Mrs. Elmo Cody was taking her eight-year-old son, Bobby, and her two nephews, three-year-old Donald and five-year-old Steve, to see Santa Claus. Mrs. Cody's car collided with a Dallas interurban car. Motorman Stephenson, the operator of the interurban car, said the old sedan driven by Mrs. Cody darted in front of him as he started across an intersection. Stephenson said he applied the brakes but could not prevent the collision. Mrs. Cody and her two nephews died instantly. Only her son survived the crash. It would be a blue Christmas for the Cody family.

December seemed to have more collisions than any other month. It could have been due to shorter days with less daylight or the distractions of the holiday season. Farmer John W. Bezdeki of West crossed in front of an interurban car on December 9, 1942. (West, Texas, is a small town between Dallas and Waco, famous for its kolaches.) Bezdeki did not make it to the other side of the tracks. The collision killed his wife and injured both of his daughters. Bezdeki was taken to the hospital, where he died from his injuries the next day.

There are numerous Texas tales of death and destruction found at the intersections of roads and rail. These are just a few of those stories. Imagine how many more occurred across the United States as America built more roads and more rail. In retrospect, it seems preposterous that these railroad crossings lacked warning lights, crossing gates or block signals.

9

THE 1914 WOODLAKE INTERURBAN WRECK

Climb the stairs behind the offices of the Interurban Railway Museum, and you will find the archives—a space filled with maps, books, photographs and railroad artifacts. Among the files is a binder of research notes that was recently rediscovered. Those notes eventually became the book *Texas Electric Railway*, by Johnnie J. Myers. At first glance, the notes, compiled by Myers, W.P. Donalson Jr. and Myers's grandfather W.T. Jacobs between 1949 and 1963, appeared to simply be a collection of descriptive information about specific interurban cars.

However, a closer look revealed something unexpected. Scattered throughout the notes were numerous comments from former employees about equipment, routes and the men who made the interurban run. Reading through the notes, references to "Marion 'Daddy' Rutledge," a crash near Wood Lake and a "letter to Bettie" jumped out. There had to be a story here, though terrible. And what an idyllic place for it to happen, at a summer retreat for families to experience relaxation and fun.

The following are the recollections of former interurban employees who would have been familiar with the people, the circumstances and the Denison & Sherman Railway cars and routes involved in the story. From these quotes, it was revealed that "Marion Rutledge" died in a crash near Woodlake around 1913 or 1914. He was reportedly the driver of car "52" and he had some relationship with "Bettie."

> *The electric cars came to Denison in 1892. I was motorman on these little electric cars until I began to work for the Denison & Sherman Ry., known*

As the cities of Denison and Sherman grew, the interurban was used not only for commuting to work but for weekend excursions as well. Woodlake Park, located between these two cities, became a popular attraction. This park provided interurban passengers with an area for activities, such as swimming, fishing, boating, picnicking and playing baseball. Other attractions included a Victorian casino, a dance hall, a penny arcade and a pavilion for theatrical productions. *Postcard courtesy of the Johnnie J. Myers Archives.*

We had mule cars first in Denison, Texas. Had twelve little street cars drawn by two mules at a time. The numbers of the cars would run from number One to Twelve. Size of cars twelve to eighteen feet long and six feet wide. Would haul 14 to 30 passengers. It was single circle line about four miles long. Began working at this when I was 15 years of age and I was a driver for mule cars. The mule car station was located at Lowe Ave. and Chestnut Street. I worked for the mule car company until the electric cars came in. No 1 and No 2 was the numbers of the electric cars – ran from 100 block Main Street to out near the Oil Mill but now at present the Kraft Cheese Plant is there. The electric cars came to Denison in 1892.

I was motorman on these little electric cars until I began to work for the Denison & Sherman Ry., known as the D.&S. Co. The Denison & Sherman Electric Railway was located at Wood Lake between Denison and Sherman, Tex. Had five cars and their numbers run from 21-22-23-24 and 25. Later on it run into Texas Traction Co. taken it over and the location was still Wood Lake.

You ask about the dummy line. It was eight miles circle track in south part of Denison. The location of the roundhouse then was two miles south of Town. Had two little steam engines, No. 1 and No 2. Very small – weigh about thousand pounds each. Had six coaches numbers run from one to six. Seat about 30 passengers. Had a ball park south of town, they would go to, but didn't have a depot or station. Just catch the dummy anywhere along the line. The locomotives they used looked like a little square box.

J. W Kidd
March 6, 1949.

An excerpt from research notes by Johnnie J. Myers for his book *Texas Electric Railway*. J.W. Kidd (March 6, 1949). *Courtesy of the Johnnie J. Myers Archives.*

as the D.&S. Co. The Denison & Sherman Electric Railway was located at Wood Lake between Denison and Sherman, Tex. Had five cars and their numbers run from 21, 22, 23, 24 and 25. Later on it run into Texas Traction Co. taken it over, and the location was still Wood Lake.

To my best knowledge, Daddy Rutledge was driving "52" at the time of the accident. Bettie has a letter Rutledge mailed her from Dallas as he left that morning—of course, she received it after his death. 21, 22, 23, 24 and 25 were the numbers of the cars used between Sherman and Denison. I'm not sure what became of them.

The first cars that ran between Denison and Sherman were nos. 21, 22, 23, 24 and 25. The 25 was torn up in a wreck in April 1910 between Woodlake and Denison. The 24 was wrecked at Woodlake in 1914, when Rutledge was killed. I don't remember when the other three cars were scrapped, but about 1917. They equipped four other cars with 4 GE 203 motors. The numbers were 188, 189, 190 and 191. They were used for plug cars between Sherman and Denison. They pulled them off during the Depression.

No, I have no pictures of the old D.&S. cars, but the numbers they carried when I was broken in on that line in 1912 was 21, 22 23 and 24, and as I remember, 24 was a little different from the other three, a little smaller and no place to carry baggage, and passengers could enter at the front end when necessary. These numbers were all changed later, I think.

E.A. Oliver of Los Fresnos was our ass't supt. For a number of years, he was motorman on north end before the south end was operating. I thought sure he would remember the no. of express car that Rutledge was operating when killed, but he let me down; also, I still am looking for someone that will back me up on the nos. of the old Denison-Sherman cars; you will note that he says that the nos. were changed, and I still say changed to 185, 186 and 187. They used three of these cars here in Waco during War #1 on Prov. Hts. And North 5th along with Trinity Hts. [cars]. *I can see that 187 with one or two trailers loaded with soldiers headed out N 5th for Camp McArthur on Prov. Hts. Headed for Rich Field.*

Now, to the details from the day of the wreck. From the *McKinney Courier-Gazette* and the *Whitewright Sun* newspapers, we find Motorman Merrell

Rutledge was involved in a head-on collision at 3:25 p.m. on a curve just south of Woodlake on July 25, 1914.

Rutledge left Sherman, heading to Denison, in a baggage and express car. It was just him and the baggageman, Mr. Nunn. As they neared the Woodlake carbarns and station and rounded the curve, a Sherman and Denison interurban car heading to the Sherman Ballpark also entered the area. It was a collision neither won.

Baggageman Nunn was uninjured, and Motorman Banks, who drove the Sherman and Denison interurban car, escaped with a sprained ankle. Mr. Rutledge, only twenty-seven or twenty-eight years of age, a six-year employee of the traction company at the time, died about two hours later. Coincidentally, almost four years prior, on May 13, 1910, these same exact cars were in another wreck just a few yards farther north. A section man by the name of Johnson was killed then. This curve then had a reputation for danger.

So, how do the facts line up with the memories of the various interurban employees? There was no "Marion" Rutledge, but there was a "Merrell/Murrell" Rutledge. The last name was correct, and the first name started

A three-car Denison and Sherman Railway train at Woodlake Park (circa 1901). The lead car is no. 24; other car numbers are not visible. All were crowded with passengers, as it was not uncommon for groups to charter two, three or even four cars for daylong excursions to the park. *Courtesy the Richard Hood Collection.*

with the same letter "M." If they called him "Daddy" Rutledge on a regular basis, not knowing his correct first name would have been quite possible. The memory of when the wreck occurred, "about 1913 or 1914," placed us in the correct date range; it specifically occurred on July 25, 1914. Rutledge was killed in the wreck, the only casualty. The numbers of the cars involved were not identified in the newspaper articles of the time. And unfortunately, the comments recorded in the research notes don't definitively clear this up. However, according to Johnnie J. Myers, the car involved was no. 25 rather than no. 52, as noted by B.C. Cook. This was probably a transposition error when recording his memories for the research notes.

But before we go—what kind of man was Murrell "Daddy" Rutledge? We don't know other than speculations that come from newspaper articles published after the wreck. The *McKinney Courier-Gazette* remarked, "He was one of the most popular employees of the company and was well liked by all."

Thanks to Hunter Herring for researching and writing this story.

10

THE MYSTERY OF JOHN H. DILLEHAY'S DEATH

While a technological marvel, the interurban was not without its share of dangers. High-voltage electricity, train track hazards and the great force the train itself could deliver were some of the threats that came with the new technology to North Texas in the early 1900s. Fortunately, situational awareness served as fine protection against the dangers brought by the railway itself. Most people were able to coexist comfortably with the mighty interurbans. Some would even arrange to meet up at the interurban tracks. John H. Dillehay, however, discovered that the dangers of trains and perilous people were entirely different matters.

Born on August 15, 1868, John H. Dillehay's rise to prominence in Collin County was practically destiny. In 1890, at the age of twenty-one he had no money, but he did have a little luck. He came under the employment of C.C. Gregory, Collin County pioneer and his future father-in-law. He worked as a farmhand for a mere fifty cents per day for four years, impressing Gregory with his hardworking character. Gregory reminisced that he "never knew a more moral, high-class, kindly and gentlemanly man," and that he felt Dillehay had an "honest desire to accomplish something in life." It appeared that Gregory's instincts were correct. In the thirty years since his farmhand days, Dillehay accomplished far more than "something." By December 1920, Dillehay had accrued $100,000, acquired 230 acres of farmland and started a wonderful family.

In fact, it was for his new family that Dillehay was away from his Parker community home on December 15, 1920. He was touring the town of

1 PLACE OF DEATH
County Collin
TEXAS STATE BOARD OF HEALTH
BUREAU OF VITAL STATISTICS
STANDARD CERTIFICATE OF DEATH
Reg. Dis. No.
Registered No. 312
B.O.V.S. FORM D
City McKinney (No. St.; Ward)
2 FULL NAME John Holland Dillehay (a) RESIDENCE. No. St. 37084
(If nonresident give city or town and State)
Length of residence in city or town where death occurred yrs. mos. ds. How long in U. S., if of foreign birth? yrs. mos. ds.

PERSONAL AND STATISTICAL PARTICULARS
3 SEX Male
4 COLOR OR RACE White
5 SINGLE, MARRIED, WIDOWED OR DIVORCED (write the word) Married
6 DATE OF BIRTH Aug 15 – 1868 (Month) (Day) (Year)
7 AGE 52 yrs. 4 mos. 1 ds.
If less than 2 years state if breast fed Yes. No. If less than 1 day hrs. mins.
8 OCCUPATION
(a) Trade, profession or particular kind of work Farmer
(b) General nature of industry, business or establishment in which employed (or employer)
9 BIRTHPLACE (State or country) Tennessee
PARENTS
10 NAME OF FATHER Tom Dillehay
11 BIRTHPLACE OF FATHER (State or country) Tennessee
12 MAIDEN NAME OF MOTHER Eliza Kemp
13 BIRTHPLACE OF MOTHER (State or country) Tennessee
14 THE ABOVE IS TRUE
(Informant) Aldridge Dillehay
(Address) McKinney Tex
15 Filed 1-4 1921 R L Davis Registrar

MEDICAL PARTICULARS
16 DATE OF DEATH Dec 16 1920 (Month) (Day) (Year)
17 I HEREBY CERTIFY, That I attended deceased from Dec 16, 1920, to Dec 16, 1920
that I last saw him alive on Dec 16, 1920
and that death occurred, on the date stated above, at 6 [illegible] m.
The CAUSE OF DEATH* was as follows:
Gun shot wounds by unknown hands
(duration) yrs. mos. ds.
Contributory (Secondary)
(duration) yrs. mos. ds.
18 Where was disease contracted if not at place of death?
Did an operation precede death? No Date of
Was there an autopsy? No
What test confirmed diagnosis?
(Signed) [illegible] Coroner M. D.
Dec 17, 1920 (Address)
*State the Disease Causing Death, or in deaths from Violent Causes, state (1) Means and Nature of Injury, and (2) whether Accidental, Suicidal, or Homicidal. (See reverse side for State Statutes.)
19 PLACE OF BURIAL OR REMOVAL [illegible] DATE OF BURIAL [illegible] 18 1920
20 UNDERTAKER [illegible] Massie ADDRESS McKinney
E. L. Steck, Austin 2887-818-100M

John H. Dillehay's death certificate. The cause of death is listed as "gunshot wounds by unknown hands." *Photograph courtesy of the Johnnie J. Myers Archives.*

McKinney in search of a new home. He was attracted by McKinney's excellent schools, hoping to provide the best education possible for his children. The Texas Electric Railway conveniently ran through McKinney, stopping at multiple stops within it and allowing residents affordable transportation to nearby towns. The day of his visit turned out to be eventful for both the city of McKinney and John Dillehay, perhaps unusually so. While Dillehay had planned to visit some houses, he participated in several other activities around the town. He saw the fire that burned down Richard Bass's home. He even considered attending the local Odd Fellows Lodge. As he explored, Dillehay stopped by multiple stores as well as the cotton mill, chatting amicably with the locals inside and acting positively cheerful.

Could Dillehay's cheerful countenance have hidden an awful truth? At around 5:20 p.m., Dillehay finally proceeded with inspecting houses. He visited the Bearden residence, a well-known vacant house that was up for sale. The visit would be one of Dillehay's last known locations before his fateful encounter. At 6:20 p.m., Dillehay's body was found, still warm, by an unlucky interurban motorman. He was flat on his back on top of a switch post at the Dowell stop of the interurban in McKinney. His right arm was draped over the main track. A bullet wound was strikingly visible on his forehead, and the gun that inflicted it lay by his left side.

Many theories emerged about how Dillehay could have met such a gruesome end. Suicide, perhaps? Dillehay had no enemies to speak of and a life insurance policy of $50,000. Could his finances have been on less solid ground than he had implied? The weapon at the scene seemed to indicate that they were. The Dillehay family owned two firearms, neither of which were the gun found at the scene. So, it appeared that Dillehay's assailant had attempted to misdirect authorities by placing the murder weapon next to him. However, the evidence was clear: he had been murdered. The reasons for his murder remain unknown. Dillehay often carried $15 on his person but was found with only $2. In 1920, $15 was the equivalent of $210 in 2022, which is quite a bit of money to be carrying around daily.

Unfortunately, the murderer was not so foolish as to loiter next to the crime scene for questioning. The police soon found themselves with no solid leads. There were rumors of a stranger accompanying Dillehay while he visited the cotton mill. There was another story of a speeding car turning into the alley where the Bearden residence was located at the time that Dillehay was inspecting the home. These hints, however, could just as well have been figments of the imagination. No faces could be pinned to these figures, let alone guilt.

A suspect was finally apprehended just a year after Dillehay's death. The Dillehay family received a letter from a law firm in Oklahoma, which stated that a certain Jerry Jones might have had something to do with the murder. Jones was shipped all the way to Texas and held on a $10,000 bail in McKinney, but no further evidence of a trial or conviction of Jones can be found. It's not clear if a trial ever took place. The circumstances of John H. Dillehay's death remain hidden; the true culprit remains a mystery as well.

Former Interurban Museum intern Austin Ng researched and wrote this story.

11
TEENAGE KILLER

Arlington Heights is a charming historic neighborhood just west of downtown Fort Worth, Texas. In the 1920s, it was considered a suburb. Residents could hop a trolley for the short ride to the downtown area. On October 17, 1920, motorman James Acord was murdered while driving an Arlington Heights streetcar to the end of its line.

Acord had just started the return trip to Fort Worth from Lake Como, a trolley park, when he was murdered. Trolley parks were created by transit companies. They were the Disney World or Six Flags amusement parks of their day. In an era when few owned automobiles, the trolley parks encouraged people to use the transit company trolleys.

It was 9:50 p.m. when a local man, D.D. Mattison, heard three shots ring out from his home. When Mattison approached the scene, he saw the streetcar jerking and sputtering before it came to a stop. Mattison found Acord slumped over his controls with a bullet wound to the back of his head.

Acord was a World War I veteran who was only twenty-seven years old. He had been employed as a motorman for a little over a year. His sister, Mrs. A.P. Waters, placed a statement in the October 23 edition of the *Fort Worth Record* newspaper: "We wish to thank our many friends for their kindness, sympathy and beautiful flower offerings at the death of our dear brother, James A. Acord."[20]

At first, there were two theories as to the circumstances of Acord's murder. The first was that he was a victim of jealousy. Detectives were sent to investigate his private life. However, his employer told the police he was steady and reliable. The second theory blamed a Black passenger seeking revenge. Arlington Heights is located next to the historically Black community of Como.[21]

His death remained a mystery for several weeks until police brought in nineteen-year-old Homer Barnes. He confessed to the armed robbery and shooting of Acord when the motorman refused to stop the train car. Barnes spent a week alone in the county jail. On October 20, Barnes waived his preliminary hearing and asked for bond. The bond was denied in Justice Maben's court.[22]

When the trial started, the State of Texas sought the death penalty. The defense made the appeal for an acquittal on the grounds that Barnes signed his confession under duress and that he suffered from "cycle insanity."

To prove the cycle insanity ailment, the defense brought Barnes's mother, Martha Barnes, to the witness stand. "What was Barnes's condition at birth and childhood, with reference to having a sound mind?" Asked the defense. "He was a premature child. He was always nervous. We had to take him out of school when he was small because he was so nervous. We had him operated on once. He was three when we decided he was not just right. Dr. R.B. Sellers, after a thirty-minute conversation with Barnes, testified that he believed him to be of sound mind."[23]

When Barnes took the stand, he admitted to being at Lake Como and that he heard the shots and saw the trolley jerk to a stop. However, he claimed that he was coerced into signing all three confessions of murder. A man named G.L. Phillips, who lived close to Lake Como, testified that Barnes gave the name E.C. Myers when he was found near the bloody streetcar. He also stated Barnes could not give a clear account of himself. In another strange twist, Barnes enlisted in the army immediately after the murder. Army recruiting officer J.C. Douglas testified that Barnes had "inquired a little oftener than the usual applicant does about being shipped."[24] On November 17, 1920, Barnes avoided the death penalty. However, he was convicted of murder and received fifty years in the Texas State Penitentiary in Huntsville.

FORT WORTH YOUTH GUILTY OF MURDER IS JURY'S VERDICT

Homer Barnes, Tried for Murder of Motorman James E. Acord, Is Given Fifty Years' Sentence.

MURDER AROUSED ALL FORT WORTH

Headline in the *Austin American*, November 18, 1920. *Photograph courtesy of Newspapers.com.*

In October 1926, Barnes escaped from Huntsville and headed for Oklahoma. In Oklahoma City, he was arrested for motor theft and white slavery. United States district attorney Roy S. Lewis had Barnes returned to Huntsville.[25] Even after this incident, Barnes walked free with a full pardon in 1934. He was one of the many convicts governor Miriam A. Ferguson released on full pardon, partial pardon or parole that year. After Barnes received his pardon, he lived out his days in the shadows, never to be heard from again.

Ferguson was the first female governor of Texas, serving two terms (1925–1927 and 1933–1935). She was the wife of ex-governor James Ferguson. Ferguson was indicted and impeached in 1917 by the Texas Senate on ten charges that including political retaliations against the University of Texas, the misappropriation of funds and receiving money from unnamed sources. Since he could not run for governor or hold any Texas office, he ran his wife's gubernatorial campaign.

During Miriam Ferguson's time in office, she granted over four thousand pardons.[26] Rumors persisted that the pardons were granted in exchange for cash paid to her husband, the disgraced, impeached James Ferguson. Jayne Y. McCallum, the secretary of state for the State of Texas, published a pamphlet shedding light on the governor's excessive pardons.[27] In 1936, Texas voters passed legislation that prevented governors from granting pardons.

12

ANARCHY IN EL PASO

El Paso is located in the far-western tip of Texas, on the northern bank of the Rio Grande. Its sister city, Ciudad Juarez, lies on the opposite bank of the river. To the north are the Franklin Mountains and New Mexico. El Paso is actually closer to the capitals of New Mexico and Arizona (Santa Fe and Phoenix) than it is to Austin, the capital of Texas.

In 1598, Spanish conquistador Don Juan de Oñate's expedition traveled north through Mexico. When they arrived at the south banks of the Rio Grande, Oñate noticed a gap between the Franklin Mountains and the Mexican Sierras. He named the area El Paso Del Norte, the "Pass of the North."

The first settlement in El Paso was founded in 1680. In 1849, the United States military established Fort Bliss, and the surrounding settlement was known as Franklin. El Paso County was established in 1850, and the community was renamed El Paso in 1852. El Paso was incorporated as a city in 1873. The first mule-drawn trolley in El Paso began service in 1881. At the turn of the century, the El Paso Electric Railway Company was formed, and in 1902, the mules were replaced by electric streetcars.

No Booze, No Problem

By 1920, the El Paso Electric Railway Company operated 103 cars and sixty-four miles of track, serving nineteen million passengers in all the

An El Paso streetcar passes the Wigwam Theatre on East San Antonio Avenue. *Photograph courtesy of the Johnnie J. Myers Archives.*

surrounding neighborhoods.[28] The company even ran a trolley across the Rio Grande into the heart of Juarez.

"Always a tourist destination because of the proximity to an 'exotic' foreign country, El Paso becomes even more attractive with the streetcars safely transporting passengers to Juárez for a drink during Prohibition."[29]

Prohibition placed a nationwide constitutional ban on the importation, production, transportation and sale of alcohol in the United States. This extraordinary situation lasted from 1920 to 1933.

As El Paso ran dry, Juarez was a quick destination for those looking for a drink. In 1921, the *El Paso Herald* reported:

> *Juarez, evidently, has no fear of Prohibition, for all the beer distributing agencies are enlarging their storage capacities, regardless of the fact that the local brewery is coming into competition with them. The Chihuahua brewery is spending $15,000.00 for the erection of a new storage house for beer. The Monterey brewery is spending half that amount to increase its space, and Ornelas Cuellar company is spending about $10,000.00 on a cold storage plant for stocks.*[30]

American distilleries also moved south of the border, such as one in Bourbon County, Kentucky. Mary Dowling moved her family's Waterfill & Frazier distillery in Bourbon County, Kentucky, to Juarez and renamed it D.W. Distillery. It partnered with Juarez businessman Antonio J. Bermudez, who later became mayor. "As soon as Prohibition hit, Waterfill & Frazier packed up and went down there," said bourbon historian Mike Veach, author of the upcoming book *Kentucky Bourbon Heritage* (University of Kentucky Press). When they left Kentucky, the Dowlings operated Waterfill & Frazier with at most $100,000 in capital, using a small column still and pot still doubler, Veach said, producing about seven barrels of bourbon a day. According to a 1937 *El Paso Herald-Post* article, the Juarez

Juarez Police to Brand Pickpockets by Short Haircuts

A New Way To Get a Drink In El Paso

PROHIBITION officers have discovered a new method of smuggling whisky over the line from Mexico. It is illustrated above. A loaf of bread is scooped out and a bottle of whisky is inserted. The work is so cleverly done that it is believed much whisky is slipped over the line in this way. It is said the scheme originated on the Canadian border and has been worked most successfully by the bootleggers up there.

Above: Headline in the *El Paso Times*, March 7, 1921. *Photograph courtesy of Newspapers.com.*

Left: During Prohibition, enterprising travelers would sneak Mexican whiskey back into El Paso, concealed inside a loaf of bread. *Courtesy of the* El Paso Herald *and Newspapers.com.*

facility stored eight thousand barrels—that's triple the family's production while operating in the United States.[31]

Saloons flourished in Juarez, and other illicit offerings were available. Gambling establishments were everywhere, prostitution was common and if alcohol was not enough, morphine and other drugs were readily available to the border-hopping visitors.

Entering Juarez, pickpockets became a huge nuisance for the American visitors. The inebriated were especially vulnerable to the quick and slippery thieves. When the Juarez police arrested a pickpocket, they would also give them a free haircut. The close-cropped heads were easily spotted on the streets and trolleys. Visitors and tourists were advised to avoid anyone with close-cropped hair.

Of course, a trolley that traveled south of the border also brought travelers back to El Paso. The trolley was one of numerous conduits along the border (from San Diego and Tijuana to McAllen and Reynosa) for smuggling whiskey back to the United States.

El Paso's Prohibition agents found enterprising smugglers using a loaf of bread to smuggle whiskey. The loaf of bread would be cut in half and hollowed out, and a bottle of whiskey would be inserted inside.[32]

Outside of Prohibition, the El Paso Electric Railway would deal with other occurrences of mayhem and criminal activity, from unruly mischievous children to the notorious trolley bandit, a highwayman with a heart and a motorman who killed a passenger.

THOSE DARN KIDS

On May 4, 1922, three-year-old Dick Guernsey had the adventure of his young life. Young Guernsey was riding the Richmond Terrace streetcar with his mother. The motorman had exited the train to turn the trolley when the youngster decided to climb up on the motorman's stool and turn on the switch. The unqualified operator promptly drove the streetcar to the end of the line and off the track, burying the front wheels into the pavement. Luckily, no one was hurt, except maybe the motorman's next paycheck.[33]

On March 23, 1923, police were able to apprehend the "Junior Bandit Gang." The gang of thirteen-year-olds preyed on El Paso motormen. One boy would climb onto the streetcar and pretend to be "defective" by twisting his hands, holding his mouth open and staring into the distance. (*Defective* was a term once used to describe people with disabilities.) At a predetermined street corner, a second member of the gang would stand on the tracks, stopping the train. The third member of the gang would then run up to the back of the car and pull the trolley pole off the electric cable. When the motorman went to the back of the car to reconnect the trolley pole, the "defective" would grab the change from the fare box.[34] The young hooligans were turned over to one Mrs. Emma Webster.

THE TROLLEY BANDIT

In 1919 and 1920, the El Paso Trolley Bandit was as elusive as a comic book villain. Like a comic book villain, he wore the same outfit for each of his crimes: a crushed soldier hat and overalls. He always struck at night, usually close to midnight. Witnesses also described the bandit as being the nervous type.

The trolley bandit's first robbery came over the Christmas holidays of 1919. His second robbery occurred just two days later, when he robbed streetcar conductor C.A. Griffin of twenty-five dollars on the Arizona Street line.

TROLLEY BANDIT ROBS MOTORMAN IN BOLD HOLDUP

Headline in the *El Paso Times*, January 18, 1920. *Photograph courtesy of Newspapers.com.*

The Trolley Bandit was not one to take time off for the holidays. At midnight on January 17, 1920, he struck for the fifth time. Motorman H.W. Sanderson was bringing his Highland Park line trolley back to the car barn when the bandit struck, robbing Sanderson of ten dollars. His only words, "Shell it out brother!"[35]

His sixth robbery occurred the following Wednesday night. The bandit robbed a streetcar operated by M.R. Scott of twenty-five dollars. Motorman Scott noticed that the gun the bandit used appeared to be an army-issued automatic.

In February, the trolley bandit had acquired a partner for his ninth robbery. The trolley bandit and his sidekick were the lone passengers on the Manhattan Heights trolley car. When the car stopped at the Texas & Pacific Railway crossing, the bandit made his move. The two men robbed Motorman C.R. King of twelve dollars. It is not known if the Trolley Bandit gave his accomplice 50 percent of the loot.

The trolley bandit did not strike again for over a month. Maybe his day job was taking up too much of his time. He robbed his next victim, Motorman B.G. Gillespie, on the eve of Saint Patrick's Day. The trolley bandit was apparently back to working solo. The incident occurred on the Highland Park line, again close to midnight.

Pointing his army-issued automatic pistol at Gillipsie's head, the bandit snarled, "Cash in....They cleaned me last night, so I'll clean you!"[36] His language implied that he might be a gambler. The bandit escaped with twenty dollars.

On April 21, the bandit struck again on another Highland Park streetcar. This time, however, the robbery would not go as smoothly as the eleven that came before. Once again, the bandit was the only passenger on the train, but he was not the only man on the train with a gun.

After robbing Motorman F.J. Darling of twelve dollars, the bandit scurried away. Darling pulled his gun and starting shooting at the bandit. However, the bandit avoided the volley of bullets, successfully pulling off his twelfth trolley robbery. The exhausted Darling reported that he had gotten a good look at the bandit's army-issued automatic pistol.

After this robbery, the trolley bandit disappeared. Did the reality of being shot at scare him off? Was he a military man stationed at nearby Fort Bliss, as his army pistol would imply? Did the El Paso crime spree end because the bandit was transferred to another fort? The case of the trolley bandit remains an unsolved mystery.

Highwayman with a Heart

Highwayman: A thief who robs travelers along a road.
Highwaymen: Thieves who rob travelers along roads and byways—not known for their compassion. But the one encountered by Motorman W.E. Bullington had a heart.

On an autumn day in 1921, Bullington was driving his regular Fort Bliss streetcar route. Just like any other day, many people entered and exited the car along its route. But one passenger suddenly lurched to the front of the car and jabbed a gun into Bullington's ribs. The rider turned highwayman demanded Bullington turn over the fares he had collected.

Bullington handed over four dollars and told the highwayman that his wife and children would have to suffer. The highwayman, startled and confused, asked Burlington what he meant.

"I mean that I'll have to make the four dollars up to the company myself and that, in the end, my family, my wife and little children, will have to suffer for this robbery. I am a married man, and I am trying to get a start in life. You are taking bread and meat out of the mouths of my children when you take this money."[37]

The highwayman took a step back and stared off into the darkness.

"Oh, hell," the highwayman said, handing the money back to the motorman. "Take back the coin, Bo, and hand it in at headquarters. After all, I'm an honest, hardworking highwayman, and I don't want any children to suffer on my account. I'll catch another car with an operator who is unmarried."[38]

Bullington knew that the company would not make him reimburse the stolen cash, but the highwayman did not. Bullington's quick thinking foiled a robbery. When he returned to the car barn, he reported the incident.

Two police officers rode the next car out. However, the highwayman never appeared. One police officer joked, "I guess he thought our motorman was married."

Self-Defense or Murder

On February 21, 1953, City Lines streetcar conductor James Lamar Webb got into a heated argument with a passenger named Francisco Dominquez. The enraged Webb struck Dominquez on the head with a trolley brake handle. The angry blow knocked Dominquez through the open doorway of the streetcar, and his skull struck the hard pavement. Dominquez lay helpless in a hospital bed for three days before dying from the head injury. Webb was charged with murder.

At the time, Webb was on parole after serving one year of a four-year conviction of rape. He claimed that Dominquez had pulled a knife on him first and that his reaction was in self-defense. Witnesses stated that they never saw Dominquez with a knife.

Webb's criminal case began on November 9, 1953. Concurrently, the family of Francisco Dominquez filed a civil case against City Lines for $100,000 in damages. The civil case was set to begin on November 16.

City Lines' attorneys latched onto the self-defense angle in their civil case with the Dominquez family. In a way, they also threw Webb under the bus, claiming his actions were not in the scope of an employee's duties. City Lines also smeared Dominguez, claiming that he was drunk and looking for trouble.

In the criminal case, the jury was quick to find Webb guilty of murder. Webb was convicted and sentenced to the state penitentiary for twenty years. A few days later, on November 17, Judge David Mulcahy postponed the Dominquez family civil case against City Lines.

Wyndham White, an attorney for City Lines, attempted to get the civil case moved out of El Paso. White waved clippings from three Juarez newspapers. "The City Lines cannot get a fair and impartial trial as a result of newspaper publicity over the criminal trial....These stories in the Juarez papers clearly show the feeling that was aroused as a result of the altercation between Webb and Dominquez and brought into the open during the trial. The Juarez papers refer to Dominquez as our compatriot and are slanted against Webb."[39]

Trolley Death Trials Are Set For November

Operator Faces Charge of Murder

Headline in the *El Paso Times*, October 23, 1953. *Photograph courtesy of Newspapers.com.*

White continued to pontificate that it would be impossible to find jurors who had not heard about the case (conveniently leaving out the fact Webb had been convicted of murder).

Joseph Dunnigan, the attorney for the Dominquez family, countered that the Juarez papers did not have strong circulation in El Paso. He also stated that the El Paso papers had covered the case honestly and without bias. Though the case was postponed, Judge David Mulcahy allowed the case to remain in El Paso.

The very next day, on November 18, Webb's attorney in the criminal case, Jack Niland, filed a motion for a new trial. Niland charged the court with not allowing a witness to testify. Tommy Powell had testified at an inquest that he had seen Dominquez with a knife. Powell quickly left town after he testified, and his inquest testimony was not allowed into evidence. Also, Niland claimed that three witnesses who spoke to Webb after the incident were not allowed to testify.

It would be the spring of 1954 before the civil suit was settled. On March 31, the attorneys for the Dominquez family and the attorneys for City Lines settled for $14,500, right after a jury had been selected for the case.

On May 18, the state court of criminal appeals sent Webb's criminal case back to El Paso to be retried. In referring to Tommy Powell's inquest testimony being denied, the court stated:

> *With this conflict in the evidence as a background, we must pass upon the alleged error of the court in refusing the appellant permission to reproduce the testimony of the witness Powell, which was taken at an inquest held in connection with the death of the deceased. There can be no question of the materiality of Powell's testimony, since the bill recites that he had testified that the deceased had an open knife in his hand, which he held in a threatening manner toward the appellant. The state does not question that Powell's testimony was taken under such circumstances as to make it admissible. The court qualified the bill, and from this, we learn that the court did not feel that it was sufficiently shown that Powell had permanently gone beyond the limits of the state or that the appellant had exercised diligence in order to secure the attendance of the witness.*
>
> *We are not impressed with the argument of the state that the appellant should be denied the right to reproduce the testimony in question because he should have availed himself of article 486a (Uniform Act to secure attendance of witnesses from without state). It was not known what state Powell was in; therefore, it could not be shown that the witness was available because of the Uniform Act being in effect in that state. Nor are we impressed with the state's contention that his plea of self-defense was amply presented by his own testimony.*[40]

The court of criminal appeals also weighed in on the witnesses who spoke to Powell after the incident.

> *Elbert M. Soniat, called by defendant, testified that he was the dispatcher for the El Paso City Lines and on duty the night of February 21, 1953, at about 7:30 in the evening; that he received a phone call from defendant sometime between 7 and 8; that defendant was very excited and was calling from the Gateway (which is at the corner of San Antonio and Stanton Streets, where the occurrence took place); that the defendant called about 3 minutes after the fight, as a result of which the deceased died. Then counsel for defendant asked the witness, "What did he (defendant) say about it (the fight) at the time," to which question the state objected, which objection was sustained by the court, and to which question the witness, if permitted to answer, would have answered as follows, "He had given him a transfer and he told him the transfer was in his pocket, to look in there, and his report was that the man said he was a man of few words and pulled a small knife out of the pocket with the blade open," and to which ruling the defendant excepted.*
>
> *It would appear that in this case, the appellant had serious trouble with one of his passengers; his first reaction would naturally be to report to his superior and to do so with all haste because the trouble had occurred on his employer's property, and they, along with him, would be held accountable, therefore. He could not telephone from the streetcar, and so he went to the nearest place where he might do so. This, together with appellant's state of excitement at the time he made the report, supplies the necessary spontaneity to authorize the introduction of the statement as part of the res gestae of the incident.*[41]

On the day before Thanksgiving in 1954, the El Paso District Attorney's Office agreed to drop the charge of murder. Webb pleaded guilty to aggravated assault and a two-year sentence after spending one year in jail.

13
TRAIN WRECK AT WATERMAN

"The Train Wreck at Waterman Song"

September, Friday Morning, this lonesome train did come.
Out into the pine woods, it was our daily run.

But on our trip returning, a tragedy occurred.
The train derailed, with four men killed, distress is what we heard.
We heard the cries in Waterman, it broke our hearts to hear.
The wailing wives and sweethearts, they faced their deepest fear.

The agony it lingered till Sunday night so late.
From the old hotel, Mr. Adams passed on through the golden gate.
The flat cars had been loaded with timber we had fell.
The engine came uncoupled, this caused them to derail.

We heard the cries in Waterman, it broke our hearts to hear.
The wailing wives and sweethearts, they faced their deepest fear.
That fateful autumn morning, the year nineteen eleven,
Four men who rode the steaming train took their final run to heaven.[42]

The lyrics of this old folk song tell the story of a dreadful event that occurred on September 1, 1911. There is nothing vague in the lyrics, like a Bob Dylan song. Similar to other folk songs, the story is laid out in simple words for simple folks.

Wreck Victim Dead.

TIMPSON, Texas, September 6.—Reports came here yesterday of the death of C. Adams at Waterman. This makes the fourth death resulting from the train wreck there last week.

Headline in the *Houston Post*, September 7, 1911. *Photograph courtesy of Newspapers.com.*

Waterman was a logging and sawmill town, located in East Texas, near the Louisiana border. East Texas is known as the "Piney Woods" and has long been associated with timber industry. In fact, "Lumberjacks" serves as the nickname and mascot for Stephen F. Austin State University in Nacogdoches, Texas. The school is well known for its school of forestry to this day.

Like other sawmill towns, Waterman only existed due to the vast forests nearby. The town was built by the Waterman Lumber Company in 1905. A rail line led from the logging camps to the mill in Waterman.

At the end of a long, backbreaking day, the train was chugging into Waterman with a load of logs and loggers. Suddenly, the brakes on the train failed, while the weight of the logs propelled the train forward. Townsfolk were startled to hear the train whistle blowing out a warning.

The locomotive crashed, hurtling logs and loggers in every direction. Some of the logs were hurled with such force that they were driven into the ground. Many of the loggers were injured, and the four loggers who died were torn to pieces. Wives and children approached the carnage apprehensively and anxious, wondering if their beloved was among the dead.

By 1913, Waterman was but a memory. Like most lumber and sawmill towns from East Texas to the Pacific Northwest, once the lumber was harvested, the companies moved on to another forest. Waterman Lumber Company moved its mill about sixty miles north to Waskom. Sustainable forestry would not come about for many decades.

Waterman lives on though through the folk song "The Train Wreck at Waterman."

14
JEFFERSON, TEXAS

Jefferson's heyday was from around 1840 to the late 1870s. During those years, Jefferson was a steamboat trading partner with Shreveport and New Orleans. Steamboats would leave New Orleans, head up the Mississippi, veer onto the Red River, navigate across Caddo Lake and head up Big Cypress Bayou to Jefferson.

During this time, Jefferson became the sixth-largest city in Texas, with a population of more than thirty thousand. The New Orleans influence is still seen in Jefferson, with the numerous wrought-iron balconies in the historic district. Jefferson also has a long history of railroad mayhem and misadventures.

In 1889, five days before Thanksgiving, Texas & Pacific freight train no. 12 was headed north to Texarkana. Brakeman W.T. Read fell between two cars and was shredded to pieces. The crew did not even know Read was missing until they reached Stall, Texas. The train backed up the five miles toward Jefferson. The train stopped by the water tower, a mile from Jefferson. There, they found the bloody, mangled pieces of Read scattered along the track for one hundred yards.

On October 19, 1938, double murderer Donald Colvin escaped from the Rusk Asylum in Longview, Texas. Colvin had killed a woman in Houston and another in Gladewater. After the second murder, he was sentenced to death. However, the ruling was overturned when Colvin was declared insane and sent to the Rusk Asylum.

The T&P Depot in Jefferson, Texas. *Conner, H.D. (train station in Jefferson, Texas), photograph, date unknown (https://texashistory.unt.edu/ark:/67531/metapth853215/, accessed December 20, 2021); University of North Texas Libraries, the Portal to Texas History, https://texashistory.unt.edu; crediting the Grace Museum.*

Colvin made his way back to his hometown of St. Louis, Missouri, but was apprehended and placed on a train back to Texas. The train made a short stop in Jefferson, and as it started up again, Colvin jumped from the train. Colvin sprained his ankle in making his escape but limped on, disappearing into the Piney Woods.

Another source of Jefferson's fame was country music star Vernon Dalhart, who wrote these immortal words.

> *'Cos he was going down a grade making ninety miles an hour,*
> *The whistle broke into a scream.*
> *He was found in the wreck with his hand on the throttle,*
> *Scalded to death by the steam.*[43]

These lyrics are from the song "Wreck of the Old '97," recorded by Jefferson, Texas native Vernon Dalhart. Recorded in 1924, the song became the first nationwide country hit record and made Dalhart a star. Although the wreck occurred in Virginia, it's appropriate that a Jefferson native recorded a song about train wrecks. The little town on the bayou has a history of dramatic train accidents both before and after Dalhart's recording.

The first major railroad accident in Jefferson took place in 1903.Two Cannonball engines, the pride of the Texas and Pacific Railroad, collided head on just outside of downtown Jefferson. The cause of the accident was found to be a mix-up of orders. Luckily, the train engineers and crewmen were able to jump to safety and avert any fatalities. It was also a stroke of good fortune that the wreck occurred outside of the downtown area, which could have caused property damage, injuries and loss of life.

Another Jefferson train wreck occurred in 1946. On November 6, Texas and Pacific Engine 661 wrecked between Jefferson and Payne, Texas. Fortunately, this wreck involved only one train, but unfortunately, there was

A group of people posing for a picture in front of a collision of T&P Cannon Balls no. 5 and no. 6 in Jefferson, Texas. *Parker Photograph (Group Photo Posing in Front of a Train Wreck), photograph, June 4, 1903; (https://texashistory.unt.edu/ark:/67531/metapth860742/, accessed December 20, 2021); University of North Texas Libraries, the Portal to Texas History, https://texashistory.unt.edu; crediting the Grace Museum.*

a fatality. The train fireman drowned in the mud by the tracks before he could be rescued. Injuries were suffered by the engineer, the brakeman and another Texas and Pacific employee who was riding in the tender's doghouse. (Doghouses gave a railroad employee shelter while monitoring the train for shifting loads or hot box problems.) The wreck and the following fifty-five-car pileup were caused by a broken rail.

Texas archaeology steward Bob Vernon, who has engaged in significant research into East Texas wrecks, shared additional history on the curve where engine 661 wrecked. It seems that Texas and Pacific Engine 600 wrecked on the same curve earlier in 1946. The daughter of the fireman was the only injury; she was engulfed by pounds of dirt that filled the cab.[44]

A year later, in 1947, Texas and Pacific Engine 907 wrecked when entering Jefferson from the east. The train entered a thirty-five-mile-per-hour curve at the rate of seventy-two miles per hour. The ensuing derailment killed Dallas fireman W.W. Darr and injured over seventy others.

T&P train no. 907 wrecked in Jefferson, Texas, due to too much speed. *Carlson, R.H. (train no. 907 wrecked in Jefferson, Texas), photograph, October 25, 1947; (https://texashistory.unt.edu/ark:/67531/metapth860259/, accessed December 20, 2021); University of North Texas Libraries, the Portal to Texas History, https://texashistory.unt.edu; crediting the Grace Museum.*

Two of the injured brought lawsuits against the Texas & Pacific Railway. Daisy Pentecost of Kilgore, Texas, suffered leg and head injuries in the wreck. Her husband, Will, sued on her behalf. Another passenger, Richard Rand from Texarkana, injured his arm, ribs, head and neck in the incident and also sued.

The lawsuits would not be settled until December 3, 1948. The Pentecosts received $6,500, and Rand received $7,500. What makes their settlement interesting is that both Will and Daisy Pentecost and Richard Rand were Black. Their case must have been very strong in a segregated Jim Crow state to win a settlement against a major American corporation.

Also in 1947, Preston Wright attempted to wreck a westbound Louisiana & Arkansas freight train. According to the *Marshall News Messenger*, "The westbound Diesel-powered L.&A. freight was brought to a halt with no damage reported about 10:30 Sunday night when it collided into two railroad crossties placed across the rails a short distance west of the junction with the Texas & Pacific Railway main line in the northwestern part of Jefferson."[45]

The engineer of the Louisiana & Arkansas freight train spotted the crossties from a distance. He was able to bring the train to a halt just as it reached the ties, preventing any major damage or injury. Wright was arrested. His motives were unknown.

15

DISORDER ON THE BORDER

The Bandit War in Texas was mainly fought in 1915 and 1916. The Bandit War was part of a larger conflict known as the Mexican Border War, the last major conflict on American soil, which lasted from 1910 to 1919. The war coincided with World War I and was one of the catalysts of the conflict was Germany. Germany aggressively encouraged Mexican attacks along the United States–Mexican border.

During this same period (1910–20), the Mexican Revolution was wreaking havoc across Mexico and spilling over into the United States border communities. The revolution was a horrific, bloody conflict between a multitude of factions. These factions included the Villistas, led by Pancho Villa.

It's also important to remember that many Tejanos (Mexican Americans in South Texas) had been displaced from their land by newly arriving Anglo Saxons. The arrival of the railroads raised land values and taxes. "Sheriffs sold three times as many parcels for tax delinquency in the decade from 1904 to 1914 as they had from 1893 to 1903; these sales almost always transferred land from Tejanos to Anglos."[46] Some Tejanos ended up working as laborers on land they had once owned. It's safe to say that south Texas was a powder keg of racism, anger and resentment.

The Bandit War was instigated by a group of Mexican rebels known as the Seditionistas. The Seditionistas wanted to start a "race war" and reclaim the border states for Mexico. The Seditionistas were never able to construct a full-scale invasion but relied on smaller raids into Texas. Many of these

Pancho Villa. *Public domain.*

Seditionistas, or bandits as the Anglos called them, raided the trains on the Texas side of the Rio Grande.

The situation also gave Anglos an excuse to shoot, lynch or commit other random acts of violence against Tejanos and Mexicans, simply by claiming their targets were "bandits."

STORM CLOUDS ON THE HORIZON

In the fall of 1914, exasperated passengers arrived in the border town of Laredo, Texas. The passengers told of the chaos and rioting in Mexico that shut down train service in Mexico City and Saltillo. Train service was delayed for several days, and the weary travelers were grateful to be back in Texas. The trouble would soon find its way north of the border.

TRAIN SERVICE SUSPENDED.

LAREDO, Tex., Sept. 6.—Passengers arriving here today from Saltillo say that, on account of the rioting and disturbances that have prevailed in Mexico City during the last week, train service between the capital and Saltillo was suspended for several days, and the first train of the week left Mexico City yesterday, bringing out a large number of passengers.

On account of the strict censorship of news, little leaks out to Mexican cities regarding the disturbances in the capital.

Storm clouds brew as train service is suspended. *From* Austin American, *September 7, 1914, courtesy of Newspapers.com.*

Saltillo, Mexico passenger depot of the National Railways of Mexico. This was a busy terminal on the Nuevo Laredo–Mexico City main line. *Courtesy of the Museum of the American Railroad, Burt C. Blanton Collection, Portal to Texas History.*

THE OLMITO INCIDENT

Burning railroad bridges was a successful guerrilla tactic of the Mexican rebels. In early August 1915, the bandits burned a bridge between Lyford and Harlingen. The bandits also slashed the telephone and telegraph wires, leaving a large area without communication,

Two days after the Seditionistas' sabotage, a work train arrived to repair the damage. The work train was backing down a switch, allowing a passenger train to pass, when bedlam ensued. Mexican snipers started firing on the work train. By the time officers arrived on the scene, the snipers had vanished into the mesquite. Luckily, no one was killed. The only one who suffered an injury was the train's engineer, who suffered a gunshot wound.

Olmito, Texas, lies just north of the border town of Brownsville. It was here, around midnight on October 15, 1915, that the gang of twenty bandits, under Luis de la Rosa, conducted a daring and bloody train robbery. By the time the escapade ended, there were five dead Americans and five more with gunshot wounds and two Mexican bandits lynched, six gunned down and four in jail.

The crafty rebels removed the rail spikes from a section of rail on the St, Louis, Brownsville and Mexico line. Next, the bandits tied a long, heavy cable to the rail and waited, secluded in the rail side brush. As the train approached, the bandits yanked the cable, pulling the rail out from under the front of the engine. The engine and first few cars of the train tumbled like toys, disengaged from the track. The front trucks of the smoker car were buried into the crossties but remained on the tracks.[47] The succeeding day's coach also remained on the tracks.

Five rebels stormed the coaches, while the other fifteen kept up a steady barrage of gunfire. The bandits shot below and above the train, keeping passengers sequestered in the cars and adding to the state of confusion.

The five rebels' first shots were aimed at the passengers in military uniforms. One soldier died from their bullet wounds, and the rebels proceeded to steal his shoes. The rebels then told all of the Mexican passengers that they would not be robbed or hurt. The rebels began robbing the other passengers on the train. The stolen booty included more shoes, jewelry and money, leaving passengers barefoot and broke but glad to be alive.

About a football field away from the chaos and carnage was an oil well camp. The oilers, awakened from their sleep, crept up close to see what was going on. Realizing they were badly outnumbered, the oilers remained hidden in the mesquite.

As part of their run to the border, some of the rebels burned a railroad trestle north of the robbery. This caused the southbound American soldiers from San Benito to leave their train and hike over a mile to the crime scene. When the soldiers finally arrived, the rebels were long gone.

A local posse lynched two of the rebels, and four more were arrested by the county sheriff. The posse caught up to the bandits in Brownsville, shooting and killing ten.[48] Did they really catch the rebels or just ten random Mexicans?

The local sheriff W.T. Vann later said that Ranger Captain W.T. Ransom was responsible:

> *Captain Ransom had* [four of the suspects] *and walked over to me and says, I am going out to kill these fellows. Are you going with me? I says no, and I don't believe you are going. He says, if you haven't got guts enough to do it, I will go myself. I says, that takes a whole lot of guts, four fellows with their hands tied behind their backs; it takes a whole lot of guts to do that.*[49]

A BLOODY THREE MONTHS

Appropriately, on Halloween in 1915, the *Austin American Sun* newspaper gave a graphic three-month report on the Bandit War.

- *July 22: Bernard Boley was killed by Mexican armed with rifle near Raymondville. Mexican escaped.*
- *July 31: Bonifacio Benivides was killed by bandits near Los Indios.*
- *August: Engineer Thompson was wounded in arm when Mexican bandits fired on work train near Sebastian.*
- *August 6: "Sonny" Huff wounded when band of Mexicans fired on automobile on Point Isabell road near San Benito.*
- *August 4: Private McGuire, Troop A, Twelfth Cavalry, was killed, and Deputy Sheriff C.A. Manahan wounded in leg during battle with Mexican bandits at Los Tulitos ranch.*
- *August 7: A.L. Austin and son, Charley Austin, executed by bandits following raid on Sebastian. Men were taken from their homes, stood up against the fence and shot. Ranger Joe Anders wounded in head shortly after in fight with bandits near Paso Real.*

- *August 7: Charley Jensen, night watchman at cotton gin in Lyford was killed by Mexican bandits who attacked him while he was making the rounds of gin at midnight.*
- *August 8: Frank Martin and George Forbes, civilians, and Corporal Mercer and Privates Reedy and Woods wounded during battle with band of seventy-five bandits at the Norias ranch house.*
- *August 10: Private Waterfield killed while on outpost at the Palm Gardens near Mercedes. Band of five Mexicans fired volley into body of soldier when he commanded them to stop.*
- *August 17: Corporal Williams killed, Lieutenant Roy C. Henry and Private Jackson wounded during battle with Mexicans at Progresso Crossing. Mexicans fired from other side.*
- *September 2: R.E. Donaldson and J.T. Smith, Americans, murdered by band of Mexican bandits at the Fresnos tract, following the firing into automobiles and burning of a railroad bridge near Barrreda the night before.*
- *September 13: Privates Anthony Kraft and Harols Foley killed and Sergeant Joseph Walsh wounded during battle with Mexicans at the Galveston ranch. Mexicans surrounded place at night and fired upon soldiers while they were asleep.*
- *September 25: Private Henry Stubblefield was killed and Captain A.V.P. Anderson wounded and Private Richard Johnson missing following battle between Mexicans and Americans at Progresso. Mexicans fired* [on] *store and attacked soldiers. Mexicans later captured confessed that Johnson was taken across the river, tortured by the Mexicans. Head severed from body and thrown in river. His ears were exhibited among Mexicans.*
- *September 26: Miss Carter, living near Harlingen, was shot in the arm by bandits who appeared at her home. She went to a well to get water, and Mexicans attempted to capture her. She had a pistol and made a fight, the Mexicans running after three shots.*
- *October 19: Dr. E.S. McCain, Engineer H.H. Kendall, Fireman B.B Woodall, Corporal Albert T. Mcbee killed. Harry J. Wallis, Corporal C.H. Layton and Private Claude J Brasher wounded when Mexican bandits wrecked then fired into a train five miles from Brownsville on the St. Louis, Brownsville and Mexico Railroad.*
- *October 21: Sergeant Shafer and Privates McConnell and Joyce killed and six private*[s] *wounded when band of seventy-five or more bandits attacked small detachment of soldiers at Ranch Ojo de Agua near Mission, five of the bandits being killed and a number wounded.*

- *October 24: Private Herman Moore received injuries from which he died the following day when in a running fight between five soldiers and twenty or more bandits, near Olmitio and within twelve miles of Brownsville.*[50]

The attacks subsided over the winter months but then resumed in March 1916.

BATTLE OF BARREDA BRIDGE

On March 15, 1916, sixteen United States soldiers were guarding the Saint Louis, Brownsville and Mexico Railroad's Barreda Bridge about fourteen miles north of Brownsville. The idle soldiers were thrust into action when they were attacked by over thirty Mexican rebels.

The *Houston Post* reported that over one hundred shots were fired but that none of the soldiers were killed or injured. It was unknown if the Seditionistas suffered any casualties.[51]

The conflict quickly deescalated when soldiers arrived to reenforce the American numbers. The soldiers were on a southbound train that was flagged down by one of the soldiers. The Mexican rebels fled to the border, and the Barreda Bridge was saved.

It was the first confrontation in the Brownsville area since the previous October.

A COWBOY WITH NO NAME

On the night of June 17, 1916, a nameless cowboy stumbled into the sheriff's office in Webb, Texas. Webb lies twenty miles north of the border town of Laredo. The dirty, decrepit cowboy told Deputy Sheriff Jack Hill that he had been captured by Mexican rebels and escaped. The battered buckaroo also told Hill that the bandits were planning to raid Webb and incinerate the railroad bridge. Hill called the *Laredo Weekly Times* one hour before midnight, sharing the information.[52]

The newspaper quickly contacted former Texas Ranger captain Tom Ross, the local United States military headquarters and the Laredo Sheriff's Department. Before the clock could strike midnight, Captain Ross was

THREE MEXICAN BANDITS KILLED AND THREE CAPTURED BY POSSE

Attempted to Burn Bridge at Webb Last Night, But Posse Was There to Meet Them.

Cowboy Who Escaped From Bandits Warned Jack Hill and He Sent Word to Laredo; Flag of Bandits Captured.

Headline in the *Laredo Weekly Times*, June 18, 1916. *Courtesy of Newspapers.com.*

leading a posse of cowboys and stockmen toward the railroad bridge in Webb.[53] However, this posse was not traveling on horseback; they were traveling the twenty miles in automobiles.

Winchesters at the ready, the posse fanned out and hid close to the bridge. In the early morning hours, three Mexican rebels approached the bridge, one lugging a can of kerosene oil. Quickly, the posse laid their sights on the bandits and ordered them to surrender. The three rebels complied, and no blood was shed as they were taken into custody. The cowboy with no name had helped save the bridge, but the troubles were far from over.

Though the rest of the rebels did not show up at the bridge, the posse felt they were close by. The posse tracked the bandits, knowing they would be heading south to cross the border. As night gave way to morning, the posse caught up with the bandits. A battle ensued, and three of the rebels were killed. The surviving rebels dispersed and retreated toward the border. The posse pursued to no avail.

MAYHEM AT THE MEDINA RIVER

Since railroad bridges were under constant threat of rebel attacks, United States soldiers were sent to protect them. Twenty miles south of San Antonio, the railroad crosses the Medina River. It was here, at the bridge, where the soldiers were attacked by Mexican rebels on June 29, 1916.

In the battle that ensued, two soldiers were injured and one rebel was taken into custody. None of the American soldiers were killed in the skirmish. It was unknown if the bandits suffered any casualties, as they dispersed southward into the night.

Chaos Reigns 1917 to 1919

By 1917, the attacks on trains and bridges had subsided while the world was in an uproar over World War I. Mexico had its own festering problems, which actually worked to the advantage of Texas's border towns.

America had attempted to maintain neutrality in World War I ever since it began in 1914. As the conflict continued to grow and intensify, the United States declared war on Germany on April 6, 1917.

Germany first tested American neutrality with the sinking of the *Lusitania* in May 1915 by a German U-Boat. Then on July 30, 1916, the Black Tom Pier in Jersey City, New Jersey, was sabotaged by German agents. The explosions killed four people and destroyed $20 million worth of military goods. Fragments from the explosion even damaged the Statue of Liberty.

This was followed by the Kingsland explosion on January 11, 1917. Kingsland is a community in Lyndhurst, New Jersey. Once again, German saboteurs were the instigators, as they totally destroyed the Canadian Car and Foundry plant. Canadian Car and Foundry had recently signed contracts with Russia and Britain to supply ammunition.

The final act that propelled the United States into World War 1 was the Zimmermann Telegram, a coded telegram intercepted by the British in

Political cartoon by John Francis Knott, 1878–1963. The cartoon first appeared in the *Dallas Morning News* on March 2, 1917. Titled "The Temptation," it depicts a satanic Germany tempting Mexico. *Courtesy of the DeGolyer Library, Southern Methodist University.*

1917. British intelligence decoded the telegram intended for the Mexican government from the German ambassador to Mexico. The telegram read:

> *We intend to begin on the first of February unrestricted submarine warfare. We shall endeavor in spite of this to keep the United States of America neutral. In the event of this not succeeding, we make Mexico a proposal of alliance on the following basis: make war together, make peace together, generous financial support and an understanding on our part that Mexico is to reconquer the lost territory in Texas, New Mexico and Arizona. The settlement in detail is left to you. You will inform the president of the above most secretly as soon as the outbreak of war with the United States of America is certain and add the suggestion that he should, on his own initiative, invite Japan to immediate adherence and at the same time mediate between Japan and ourselves. Please call the president's attention to the fact that the ruthless employment of our submarines now offers the prospect of compelling England in a few months to make peace.*
> *Signed, ZIMMERMANN*[54]

This was not the first time Germany had encouraged Mexico to attack the United States. The Zimmermann Telegram was, however, the country's most audacious call. On March 3, 1917, the German foreign secretary Arthur Zimmermann confessed to the document's authenticity.

In Mexico City, a military commission was assigned to study the plan's feasibility by Mexican president Venustiano Carranza. The commission reported that the plan was not feasible for a multitude of reasons, namely that Mexico was in the middle of a long, blood-soaked civil war (the Mexican Revolution).

AN ASTOUNDING PLOT REVEALED

It Was a German-Japanese-Mexican Alliance.

ATTACK ON THE U. S.

Mexico Was to Get Texas, Arizona and N. M.

BERNSTORFF IN DEAL

Pres. Wilson Had Knowledge of the Plot Before Relations Were Broken Off.

Headline in the *Arkansas City Daily Traveler*, March 1, 1917. *Courtesy of Newspapers.com.*

16
THE VALLEY OF DEATH

The Lower Rio Grande Valley is located at the southern tip of Texas. The area includes the border towns of Brownsville, Texas, Matamoros, Mexico, and McAllen, Texas, and Reynosa, Mexico. The bilingual region is known to locals as the "Valley." On March 14, 1940, it would become the valley of death.

The no. 113 westbound Missouri-Pacific passenger train was headed to Mission, Texas. The railway ran parallel to the Rio Grande Valley Highway. At the same time, a truck loaded with fruit pickers was traveling on the highway. The truck turned onto Tower Road, which intersects with the railroad tracks, and, unfortunately, was rammed by the passenger train.

Twenty-seven people, mostly Tejanos, were killed in the crash. It was reported that bodies were scattered along the track for five hundred feet.

Brad Smith of the *Brownsville Herald* newspaper was one of the first on the scene.

> *Horrified, I stopped to count the bodies. Within a moment or two, I counted 18 bodies. Only four of them moved. One man was half lying near the west edge of Tower Road. He was vomiting. I ran to him, and he collapsed. There was nothing I could do for him. A few feet away, a man lay on his face. His clothing was ripped from the back of his body, and red welts appeared across his back. He moved a moment or two, then lay still.*
>
> *I saw arms, legs, bodies torn into bits. I saw two arms lying between the rails. I stopped short, almost amazed, as two priests, seemingly appearing*

Bodies Strewn Along Track for 500 Feet By Impact of Blow

Extra! The Brownsville Herald Extra!

FORTY-EIGHTH YEAR—No. 222 BROWNSVILLE, TEXAS, THURSDAY, MARCH 14, 1940 ★★★★ 5c A COPY

23 VALLEY PERSONS KILLED AS TRAIN HITS FRUIT TRUCK

VICTIMS ARE MEXICANS; 17 ARE INJURED

Passenger Train Splinters Truck Loaded with Citrus Pickers.

DEATH TOLL VERIFIED

Ambulances from Five Towns Rush Victims to Hospitals and Morgues.

Top: Headline in the *Brownsville Herald*, March 14, 1940. *Courtesy of Newspapers.com.*

Middle: Headline in the *Brownsville Herald*, March 14, 1940. *Courtesy of Newspapers.com.*

Left: Headline in the *Victoria Advocate*, March 14, 1940. *Courtesy of Newspapers.com.*

> *from thin air, stepped quietly but swiftly into the scene of horror and began their last ministrations.*
>
> *Most terrifyingly significant throughout those awful first minutes was the fact that so few of the victims—so very few—even so much as moved. It was death on a scale I never before had witnessed in this rich, peaceful farming country bordering the Rio Grande.*[55]

The driver of the truck, Jose Ramon, was identified by matching his severed arm to his body. John Boeye was traveling in an automobile on the adjacent highway about a half mile behind the train. Boeye told the *Valley Morning Star* newspaper, "Bodies were still rolling along the right-of-way, and the train was still moving when they arrived."[56]

Spectators were horrified at the severed, mutilated body parts scattered along the rail line. It was the greatest disaster in the Valley since the 1933 Cuba-Brownsville Hurricane.

17
PORTER PERILS

The most influential Black man in America for the hundred years following the Civil War was a figure no one knew. He was not the educator Booker T. Washington or the sociologist W.E.B. Dubois, although both were inspired by him. He was the one Black man to appear in more movies than Harry Belafonte or Sidney Poitier. He discovered the North Pole alongside Admiral Peary and helped give birth to the blues. He launched the Montgomery bus boycott that sparked the civil rights movement—and tapped Martin Luther King Jr. to lead both. The most influential Black man in America was the Pullman Porter.[57]

After the Civil War, many freedmen (formerly enslaved people) went to work for George Pullman's Pullman Company. Pullman had created a railroad sleeping car and needed workers to serve his passengers. The porters carried bags, shined shoes, served customers and maintained sleeping quarters. Pullman also knew the freedmen would work for cheap.

Other freedmen headed to Texas and worked the cattle drives that headed north to the railroad. It's estimated that five thousand Black cowboys worked cattle in the trail drive era between 1866 and 1890. Cattle drives were dangerous, and the cowboys had to rely on each other. It created an environment in which a man was judged by his skills and work ethic, not his skin color.

As the railroads made their way into the Lone Star State, the cattle drive era came to an end. However, the railroads offered new opportunities for the Black cowboys.

An 1894 Pullman dining car poster advertisement featuring a Pullman porter on the right. *Public domain.*

A 1960 photograph of Santa Fe's *Super Chief*, daily all-Pullman streamliner operated between Chicago and Los Angeles. *Courtesy of the Museum of the American Railroad, Burt C. Blanton Collection, Portal to Texas History.*

At the dawn of the twentieth century, Black Americans, including many of the Black cowboys, found work on the Southern Railway as train assistants. The Pullman Company, owned by George Pullman, provided each "Pullman porter" with a black uniform, and they received training in a code of conduct that would be nonthreatening to the company's wealthy white passengers. While being a porter proved to be a step up from working on the range, discriminatory practices on the part of white passengers continued to plague the workers.[58]

TEXARKANA

It was almost midnight on September 15, 1904, when Porter Tom Jones was killed by a single .45-caliber bullet. The bullet entered behind his left ear and exited at his right temple, killing him instantly.

Earlier, two train-hoppers had hitched a ride near Hope, Arkansas. One tried to hide in a closet but was discovered by the conductor and forced to pay his fare. The second man was discovered by Jones, and the two men started to fight. The train's brakeman broke up the fight as the train arrived in Texarkana.

The train-hopper pulled a gun and shot Jones through the head. The train-hopper ran from the train, disappearing into the night.

Dead Men Tell No Tales

Early on the morning of April 26, 1907, Pullman porter G.D. Ballard was shot to death at Fort Worth's Texas and Pacific roundhouse. His killer, J.R. Yates, was employed by the Pullman company as a foreman of the car-cleaning crew.

Yates claimed that he killed Ballard in self-defense. Yates's story was that he had to reprimand Ballard for not cleaning up the railcar properly. This, according to Yates, angered Ballard. He said Ballard had a knife, grabbed a hammer and moved toward him. Yates said he warned Ballard twice to stop or that he would kill him. He then proceeded to shoot Ballard twice through his right side, the fatal bullet ripping through Ballard's heart. After the shooting, Yates turned himself in to the authorities. Ballard was found with a knife in his hand, but was it planted?

A grand jury indicted Yates for the killing. However, due to the witnesses, the case was dismissed on January 18, 1911, almost four years after the killing. The truth of what actually happened died with Porter Ballard.

Six years later, the *Fort Worth Star Telegram* was able to shine a light on what kind of person J.R. Yates really was. Wherever Yates traveled, death traveled with him.

- 1900: As a night manager in Lancaster, he killed Dick Whitworth.
- 1907: He shot and killed Pullman porter Ballard in Fort Worth.
- 1908: He struck a man in the head with an iron pipe. The unnamed man became mentally incapacitated and could not testify.
- 1911: He shot and killed accountant Clyde Styers at his desk in Dallas.
- 1913: He shot the son of R.H. Orr in the foot.

- 1916: He shot and killed safeblower Tex Wallace. Wallace had an abundance of charges against him and once claimed that Yates was out to shut him up.

Amazingly, Yates never served time for these crimes and was actually acquitted twice. Also, he was able to move from job to job easily.

In 1917, Yates received his comeuppance. After murdering Fort Worth police commissioner Ed Parsley, Yates barricaded himself in the commissioner's office. The Fort Worth police put the standoff to an end quickly. Yates's wicked life would end with his bullet-riddled body lying in a pool of his own blood.[59]

For Walking on the Sidewalk

In 1909, on an April afternoon in Brownsville, Texas, a policeman gunned down Pullman porter Brewer. Officer Daugherty said that Brewer was taking up too much space on the sidewalk. When a group of white women came walking toward them, Daugherty told Brewer to get off the sidewalk. Brewer refused to move.

Daugherty attempted to arrest Brewer. Brewer made a run for it, and Daugherty shot him in the back. In a statement, Daugherty said Brewer ran with his hand on his hip pocket. Regardless, a man was dead just for standing on the sidewalk.

These are just a few examples of the violence meted out toward Pullman porters, workers in a profession that was considered the best job in the Black community and the worst job on the train.

Not all porters were men. In 1943, Maggie Hudson was one of the first Black women hired as a "porterette." Maggie received the same pay rate as the male porters. Maggie retired in 1979 after thirty-six years of service. *Photograph courtesy of the Plano Conservancy.*

18

JUNE 10, 1907

BLAME IT ON THE MOON

The day June 10, 1907, seemed like just another date on a dusty, old calendar. However, this unassuming day would turn out to be bizarre, newsworthy and offbeat for Texas railroads all across the state. Was this just coincidence? There was a new moon that day. Superstitions say a new moon is a time for new beginnings. However, folklore also conveys the dangers of traveling during a new moon. It certainly proved to be a day of mayhem for Texas railroads. Did the new moon have some invisible influence on the day's circumstances? Whatever the reason, it was a strange day of accidents, justice and oddities in the Texas railroad business. The following incidents all occurred on Monday, June 10, 1907.

GUILTY

In Palestine, Texas, Monk Dudley pleaded guilty to assaulting Pacific Express messenger Winsley Womack. On March 20, Womack was working on International & Great Northern Railroad train no. 4. The train had left Houston and was heading north. Just south of Palestine, near Elkhart, Womack was attacked by Dudley.

Dudley pistol-whipped Womack, knocking him out. As the train rolled down the tracks at forty miles per hour, Dudley threw Womack's unconscious

GETS 20 YEARS

Monk Dudley Pleads Guilty to Train Robbery at Palestine.

THREW MESSENGER OFF CAR

Headline in the *Houston Post*, June 11, 1907. *Courtesy of Newspapers.com.*

body off the train. Womack was lucky to survive. Dudley proceeded to rob the express car's safe. At his trial on June 10, 1907, Dudley was sentenced to twenty years for his crimes.

A BIG NEWS DAY FOR LITTLE ELKHART, TEXAS

STATION AGENT ROBBED.

Masked Men Get Considerable Sum of Cash at Elkhart.

Headline in the *Austin American Statesman*, June 11, 1907. *Courtesy of Newspapers.com.*

One the same day that Monk Dudley was convicted for the crimes he committed on a train near Elkhart, two masked bandits were committing crime on another train near Elkhart.

The masked men pulled pistols on railroad agent T.J. Lawrence, just as the northbound train pulled out of the station. Fearful for his life and staring at the end of two barrels, Lawrence handed over $190. Like comic book villains, the masked bandits disappeared from the train before it got up to speed.

MAYHEM IN MARLIN

Marlin, Texas, is located about thirty miles southeast of Waco. On a sleepy morning in Marlin, two freight trains collided around 10:00 a.m.

The rail yard's little switch engine was backing a long line of loaded coal cars around a curve in the track. As the coal cars retreated in a southerly direction, the switch engine crew could not see around the curve in front of the switch engine. Also, their focus was on the cars behind them, not what was in front of them.

At the same time, an International and Great Northern freight train was heading south on the same track. The freight train had two empty boxcars and twenty-five loaded boxcars. The freight train's engineer stated he could not see the switch engine and coal cars around the curve of the track.

In the ensuing mashup, the two empty boxcars flew off the track, landing in the street, demolished. Also, one boxcar ended up halfway off the track.

Coal flew like popcorn popping, and both engines were damaged. Luckily, both trains were traveling slowly, and no one was killed.

CUT IN TWO IN THE TWIN CITY

The Northeast Texas town of Texarkana shares a border with its twin city Texarkana, Arkansas. This is where Adele Hains lived close to the railroad tracks. Will Driver, a local drayman (a person who delivered beer for a brewery) came by to call on Adele. Soon, their conversations turned argumentative. The enraged Driver pulled out a razor, slashing Adele across her face.

Adele swooned from the attack. Petrified that he had killed her, Driver began to run. In his frantic escape, he did not notice the approaching Iron Mountain switch engine as he ran across the tracks. The switch engine's wheels rolled over Driver like a pizza slicer, literally cutting his body in two pieces. Adele Hains survived the razor attack.

FIRE IN FORT HANCOCK

Fort Hancock is located about fifty miles southeast of El Paso on the Rio Grande. An El Paso–bound Texas & Pacific freight train was delayed here when a boxcar derailed. The boxcar's journal had caught fire and burned out, causing the car to jump the tracks. (A journal was a shaft rotating in a hole. On a boxcar, the journal bearing was the plain bearing once used at the axle ends of railroad wheels, enclosed by journal boxes. Oil could leak from the boxes, causing fires that, if not detected, could burn down an entire boxcar.)

The boxcar was moved back on the tracks. The car was the pushed onto a siding, allowing the freight train to proceed to El Paso.

STRIKING OIL IN EL PASO

Meanwhile, in El Paso, workers were dealing with an oil-flooded railroad yard. The constant rolling of heavy trains caused the underground feeder pipe to burst. The underground feeder led to the oil tanks, which, in turn, connected to the engine tanks.

A small leak in a joint led to the pipe bursting and the oil boiling out of the ground. No one was hurt. However, the surrounding area looked like the aftermath of a Texas oil well gusher.

An El Paso Burial

R.W. McClean of Chihuahua, Mexico, arrived in El Paso on Sunday night, June 9. A successful proprietor of Chihuahua's Robinson House, he did not have business in El Paso. He was in town to make burial arrangements for his good friend Gilbert C. Simpson.

The previous Thursday, Simson had arrived in El Paso on the Mexican Central Railroad critically ill. Simpson, a native of Glasgow, Scotland, had lived in Chihuahua for over ten years, working in the mining business.

The stricken Simpson was taken from the train to the Sheldon Hotel. There, a doctor examined Simpson and found Simpson's lungs were hemorrhaging. By Friday afternoon, Simpson, only thirty years old, was dead.

McClean, also a native of Glasgow, Scotland, was notified of Simpson's death and headed north to El Paso. On Monday, June 10, McClean made funeral arrangements for Simpson, and he told the *El Paso Herald*:

> *At the Orndorff Hotel, where he is a guest, Mr. McClean said that the news of the death of his friend shocked him, as when Mr. Simpson left Chihuahua for El Paso, he was apparently in fairly good health. The object of Mr. Simpson's visit to El Paso was to purchase an outfit for an independent assay office, which he expected to open in Chihuahua. Before leaving Chihuahua, Mr. McClean loaned Mr. Simpson $2,000, and when he arrived in El Paso, he had only $900 Mexican and $26 gold. The suspicion is that when he was taken ill on the train, he was robbed.*[60]

Simpson did have a history of heart trouble, but the lung damage was unrelated. Was Simpson poisoned and robbed on the train between Chihuahua and El Paso? If so, why did he still have $900 in Mexican money and $26 in gold? Did he become entangled with a few shady characters? Those answers are lost to history.

EAST OF EDGEWOOD

FAST EXPRESS IN THE DITCH

REPORT FIRST RECEIVED MORE SERIOUS THAN TRUTH.

Entire Train Derailed Near Edgewood and Bridge Superintendent of Texas and Pacific Thrown Through Window.

Headline in the *Austin American Statesman*, June 11, 1907. *Courtesy of Newspapers.com.*

The Texas & Pacific's Cannon Ball passenger train had left Dallas and was bearing toward its final destination in St. Louis. About seventy miles east of Dallas, disaster struck. The incident happened a little before noon at mile marker 157, three miles east of Edgewood, Texas.

Approaching the trestle at Mills Creek, the engine made it safely across the bridge. Before making it across the bridge, somehow, the tender became disengaged from the tracks. (A tender was a railcar behind the locomotive that carried the steam engine's fuel, either wood or coal and water.) The tender, and the cars coupled to it, tumbled into the muddy creek below.

Superintendent E. Lowery, who worked for the Texas & Pacific Railway as a bridge superintendent, was riding on the train. Lowery was propelled headfirst through a window and landed in the swampy, muddy creek, unconscious. Lowery nearly drowned in the mud before he was rescued.

Of the one hundred passengers and railroad employees on board, twenty-three were injured in the wreck. Many were injured trying to climb out of the mangled passenger cars. There were three postal clerks, two porters (one knocked unconscious) and a cook injured in the crash. The worst injury was a head injury suffered by postal clerk Frank Peders. Once out of the cars, the startled, confused passengers and employees had to deal with the mud. Recent rains had made the soggy, swampy creek even more treacherous. The uninjured helped the injured escape the muddy hell hole. Eventually, everyone made it to high ground, covered in mud, dazed, exhausted and grateful to be alive.

How did the wreck happen? Investigators, led by Texas & Pacific general manager L.S. Thorne, could not find the trucks that should have been on the tender. (Trucks are the chassis frame under a railcar that contain the wheels, axles and bearings.) Trucks were known to catch fire easily. If not snuffed out quickly, these fires would burn up entire railcars. Did the trucks explode from the heat of fire? The wreck had torn up almost 550 feet of rail. There was evidence of some faulty railroad ties. Were the ties already damaged, or were they damaged by the wreck?

The *Austin American Statesmen* reported, "General Manager L.S. Thorne returned tonight from the scene of the wreck. He says he was pleasantly surprised to find the wreck a much less serious affair than he expected. None of the injured had broken bones, and none was seriously enough hurt to have to be sent to the hospital."[61]

From Texarkana to El Paso, June 10, 1907, was a day of accidents, crashes, convictions, death and robbery on Texas railroads.

19

FAMILIES SHATTERED IN PLANO

Plano, Texas, is home to the Interurban Railway Museum. The museum shares the history of the Texas Electric Railway, which ran from 1908 to 1948. The Texas Electric Railway gave families an easy, convenient way to travel to Dallas, Fort Worth, Waco and other cities in North Texas. For some families, the Texas Electric Railway was a haunting reminder of families shattered.

DEATH CLOSE TO HOME

We have all heard the saying "most accidents happen close to home." For the Hughston family, death was just two doors down. On a Saturday evening in the summer of 1913, Arch Hughston was taking his family out for an evening drive. The passengers in the car included his wife, Emma, their four children, Oliver (thirteen), Lucile (eleven), Dick (seven) and little A.T. (three), and the six-month-old son of Mr. and Mrs. J.R. Dickerson, neighbors of the Hughston family.

The Hughston home was only 150 feet from the interurban railway tracks, and their trip turned out to be a short one. As Arch headed east, he did not hear the southbound train's whistle. Because of trees and homes that blocked his vision, he did not see the train until it was too late. Arch jerked the steering wheel to the right but was struck by the train.

On impact, Emma Hughston was ejected from the car, landing on the tracks in front of the train. The steel wheels rolled over her body, decapitating her and severing her legs. The head and torso were carried one hundred yards down the track, while her legs were left trackside. When the train finally stopped, the head and torso had to be pulled from underneath the train, with the torso wound around the rear axle.

Three-year-old A.T. was also hurled underneath the car. His small body was shredded, his limbs hanging by pieces of skin and skull fractured. Amazingly, the other passengers and Arch Hughston survived the horrible crash.

HELPLESSLY WATCHING

In March 1921, the Sorenson family was returning to their home in Harlan, Iowa. The Sorensons had spent the winter in San Benito, Texas, with their relative Harry Sorenson. The Sorensons were traveling in two automobiles.

Traveling north through Texas, Clarence Sorenson was driving the lead automobile with his nephew, Oliver Jr. (nine) and his nieces Edna May (five) and Leslie (one). Mr. and Mrs. Oliver Sorenson, the parents of the three children, followed in a second automobile.

As the family caravan entered Plano, their route passed over the parallel Houston & Texas Central Railroad and Texas Electric Railway tracks. Clarence safely passed over the Houston & Texas Central tracks, and then tragedy struck.

Oliver Sorenson and his wife watched helplessly as Clarence's auto was slammed by an oncoming Texas Electric Railway Interurban car. The impact destroyed the automobile on contact and took out the air brakes on the interurban car. As the runaway interurban car traveled down the tracks, it left pieces of the automobile scattered along the railroad. All three of the children were killed, but somehow, Clarence survived.

Details of Tragedy That Cost Three Lives Near Plano

Father Gives Graphic Picture Of Death Crash

Little Bodies Horribly Mangled By Electric Car

Headline in the *Courier Gazette*, March 31, 1921. *Courtesy of Newspapers.com.*

The motorman on the interurban car claimed he laid heavy on the train whistle. Clarence claimed he never heard the whistle. Oliver Sorenson described the bloody scene to the *Courier Gazette* newspaper:

> *The baby, one year old, was the first one I came to. I picked it up in my arms and found a bruise on one side of its head. I placed my hand over its heart and found that my baby was dead. I laid it down and ran to my little girl, five years old, some distance from the baby. I picked her up, but the body fell to pieces. Her body was cut in two at the waist. I laid her down and went to my oldest boy. He, too, was dead, the back of his head having been torn away.*
>
> *Finding that my three babies had been killed outright, I thought of my brother, who was dragged in the wrecked auto by the interurban for quite a distance and, half-crazed by what I had already seen, I ran to his assistance. By the time I arrived where the car was brought to a standstill, I saw my brother extricate himself from the wreckage and come toward me. My wife and I were only a few yards away and saw it all—saw our children killed and their little bodies mangled, and yet we were as helpless to save them as if we had been a thousand miles away. I'll tell you, it was terrible.*[62]

In a cruel twist, Clarence was taken to the hospital in McKinney, carried by the same interurban car that had instigated the carnage.

She Never Arrived

In March 1946, Mrs. E.N. Dinwiddie of Plano was anticipating the arrival of her daughter. Her daughter, Mrs. James S. Keller, and Mrs. Keller's stepmother were driving up from Dallas. One mile south of Plano, Mrs. Keller attempted to drive over the Aldridge Crossing.

As the automobile crossed the tracks, it was struck by a Texas Electric Railway Bluebonnet Special. Mrs. Keller was killed on impact, with one arm cut off and her body almost torn in two pieces. Somehow, her stepmother survived.

The next time Mrs. Dinwiddie saw her daughter was at the funeral.

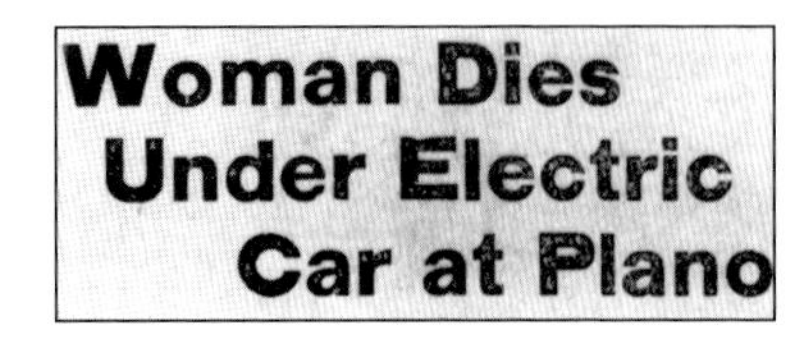
Woman Dies Under Electric Car at Plano

Headline in the *Corsicana Semi-Weekly Light*, March 15, 1946. *Courtesy of Newspapers.com.*

20
THE COST OF COST CUTTING

As part of an effort to cut costs during the Depression, the Texas Electric Railway (TERY) resorted to the consolidation of the roles of workers. Interurban cars, previously operated by a motorman and conductor, changed to a one-person operation. The motorman then performed his original role, plus that of the conductor. Along with the conductor's departure, so went the extra degree of caution that had guaranteed the TERY's impeccable safety record of decades prior. These changes are likely what caught up to the veteran interurban system in its final years.

Between 1946 and 1948, six Texas Electric Railway interurban cars were involved in three crashes, resulting in 107 passenger injuries. With ridership already on the decline after World War II, these crashes devastated the TERY.

It was a foggy morning on March 14, 1946, when car 317 lost power north of Lancaster, Texas. Even though railroad fuses (flares) were lit, car 317 was rear-ended by car 326.[63] No one was killed, but twenty-seven passengers were injured.

Six months later, the newly repaired car 317 had another accident. On September 29, 1946, car 317 was involved in a head-on collision with car 308. The crash site was between Denison and Woodlake, a heavily wooded area where the track held numerous blind curves.[64] Again, no one was killed, but between the two cars, thirty passengers were injured. Both Cars had to be rebuilt, costing Texas Electric Railway even more money.

INTERURBAN BETWEEN DALLAS, WACO CRASH, INJURING 27 PERSONS

Headline in the *Corsicana Semi-Weekly Light*, March 15, 1946. *Courtesy of Newspapers.com.*

Two Texas Electric Interurban Railway Company passenger motors collide with each other on the tracks as a group of onlookers and rescuers mill about (April 10, 1948). *Photograph courtesy of the Johnnie J. Myers Archives.*

The third crash, the worst of the three, occurred on April 10, 1948. Southbound car 365 (train 3) and northbound car 366 (train 6) ran into each other head on, injuring forty-nine passengers as well as a motorman. The motorman from car 366, Z.L. Lowe, had been told to meet car 365 at the Kirkland siding, just north of Dallas. (A siding is a short length of railroad track that allows for two trains on the same line to pass each other.) The motorman misunderstood the directions and kept rolling through the siding, right into car 365.

J.W. Davis, the motorman from car 365, said, "I looked up and saw the other car. There wasn't much time to do anything. I threw on the brakes and then stepped back out of the front vestibule, into the car. That probably saved me from a great deal more than a broken shoulder."[65]

Mrs. Lillian Christie was a passenger on car 365 and gave a firsthand account of the accident, "I saw our motorman turn and start for us. He bumped into me, and I was knocked down. Then the whole front of the car fell in on me. I was not unconscious."[66]

Opposite: Interior of Texas Electric Railway passenger motor 365, wrecked in an April 10, 1948 accident. It shows the chaos created by the impact of the wreck. *Photograph courtesy of the Johnnie J. Myers Archives.*

This page, top: Another interior picture of wrecked Texas Electric Railway passenger motor 365. *Photograph courtesy of the Johnnie J. Myers Archives.*

This page, bottom: At about 7:40 a.m. on April 10, 1948, southbound train 3 (car 365) met northbound train 6 (car 366) north of the Kirkland siding, near the present-day Royal Lane in north Dallas, with the result of a crash between the two cars. There were no fatalities, but thirty passengers and the 365's motorman were injured. For Texas Electric, it was the straw that broke the camel's back. The accident, the third serious one in little more than two years, led management to the "abandonment decision." *Photograph courtesy of the Johnnie J. Myers Archives.*

This disastrous crash was the death knell for the Texas Electric Railway. The Interstate Commerce Commission, in response to the crashes, mandated an update to all rail crossing signals. The update was beyond what the TERY could afford. Impending lawsuits from those injured all but guaranteed rising expenses for the railway. Of course, passengers' wariness of the trains' safety hurt ridership as well. Low on funds—and with them only getting lower—the stockholders decided that the closure of the line was the only option. Ten days after the disaster, Texas stockholders met and decided to shut down the interurban lines.

After over forty years of service, rising competition from cars and increasing questions about safety sent the Texas Electric Railway out with the old on December 31, 1948.

21
MISCELLANEOUS MAYHEM

Asleep on the Tracks

Even some employees of the Texas Electric Railway lacked respect for the dangers of new technology. On a summer day in 1915, a seventeen-year-old watchman named Howard Rogers fell asleep next to the interurban tracks near Waco. He did not hear the incoming train car and woke up too late. The train car's cowcatcher decapitated Rogers, killing him swiftly. The train car, its motorman and its passengers escaped unscathed. Whether Rogers was uninformed of the dangers, ignoring the risks or fell asleep next to the tracks by pure accident, it was a fatal mistake.

Current Conductor

Almost all interurban cars, trolleys and streetcars were powered by electricity, exposing railroad employees to an unfamiliar technology and a new set of potential hazards.

The cars were connected to an overhead electrical line by a trolley pole, which powered the motor, lights and heaters on the car. The poles had a swivel base and were spring-loaded to keep in constant contact with the overhead cable. This kept the pole engaged while the car navigated curves,

Using an interactive exhibit at the Interurban Railway Museum, Docent Aaron Casperson helps a visitor place the trolley pole back on the power line. *Photograph by Charlotte Loontiens.*

provided extra height when crossing other rail lines and offered the option of passing under a lowered line in a tunnel or underpass.

A trolley pole rope was connected to the far end of the pole and allowed the motorman to engage or disengage the electric cable without being electrocuted. The fibers of the rope were nonconductive, keeping the motorman safe from electrical shock, burns or death.

NEW NUMBER SAME RESULT

Car 306, a brand-new, 1,200-volt, "local type," two-man passenger coach motor was bought from the St. Louis Car Company by the Southern Traction Company in 1913. In 1917, the car was passed on to the Texas Electric Railway Company, bearing the same number. To reduce costs, it was converted to a one-man car around 1932. Passenger steps at both of the back doors were removed, and the passenger entrance was rerouted through a folding front door, just to the right of the motorman.

In many places, this conversion had to be approved by a municipal ordinance. While it did help companies save money by reducing the number of crewmen required for operation, one-man trolley conversions were considered unsafe practice and, as a consequence, prompted resistance from motormen and conductors alike. If employees were involved in a union, any talk of a one-man conversion would be challenged, regardless of how difficult a financial situation the company was in, often resulting in strikes or other industrial action.

Six years later, car 306 was involved in a wreck and subsequently put into storage. The *Dallas Morning News* reported:

> *A Waco-bound Texas Electric interurban and trailing freight car, driven Lisbon, and rammed into a gasoline car that had been waiting at a siding (thankfully, no explosion occurred upon impact). Samuel B. Mathis, the conductor, was in the freight car when the vehicles made contact. He was "cut on the face and bleeding severely" but declined medical aid in favor of helping those with worse injuries.*

Ultimately, twelve hurt passengers were sent to Baylor Hospital for treatment. The *Dallas Morning News* reported that the accident caused a flurry

Bluebonnet 306, southbound on the Dallas-Waco line, hit what was reported to be an open switch and "hurtled" into a gasoline car parked on a siding near the Fernwood Avenue crossing at Lisbon, in the south Oak Cliff outskirts of Dallas. The gasoline car did not explode. Following the accident, more than one thousand people crowded the streets leading to the scene. *Photograph courtesy of the Johnnie J. Myers Archives.*

of onlookers to gather around, with "more than a thousand persons [crowding the] streets leading to the scene....Twice at the scene of the accident, working newspapermen were ordered to leave." The article even revealed details of the crash, as told by the passengers who experienced it firsthand.

Car 306 sat, locked away from the public for four years, before it was converted to six hundred volts and renumbered to 366. Unfortunately, in its new life as car 366, it was again involved in an accident, which finally put it out of commission for good in 1948.

"In 1938, no. 306 was involved in a rear-end collision with a preceding train. She was carrying a trail car at the time and thus was damaged at both ends. Was placed in storage until April [19]42 when it was equipped with 600 volts," said William E. Sturgis, a mechanic and electrician for the Texas Electric Railway in the 1930s and 1940s.

No Cash and Straight to Jail

Not every robbery resulted in decent profits or clean getaways. Notorious criminal Oscar Lafferty had been in and out of prison until he was caught robbing the interurban station in Red Oak, Texas, with local criminals Fred Mace and Miss Bobbie Hines in 1932. Disguising Hines as a man, the trio raided the station, gaining no money but an assortment of items: two cartons of bullets, a ticket box, two pillows, a tube of shaving cream, a can of shoe polish and a can of powder. Oscar pleaded guilty before a judge in 1933 and received a life sentence for his repeat offenses.

Only Human

Fred Owens worked for the railroad in Gainesville, Texas. On February 28, 1908, he came off his run at 3:00 p.m. Like every other day, Owens headed to his mother's house, where he lived.

He washed up, had something to eat and told his mother he felt a little tired. Owens headed to his bedroom for a rest. A few minutes later, Owens's mother heard a gunshot. She ran to Owens's bedroom and found him dead as his bed became soaked with blood.

Owens had shot himself above his right temple, and the bullet went straight through his skull, exiting by his left ear. Fred Owens was only thirty-four years old, and the reason he took his own life remains a mystery.[67]

On September 1, 1940, Motorman Sam Dixon was only twenty-five years old when he was found dead in his home. Dixon had died from a self-inflicted gunshot to his heart. Again, no one knew the reason why.[68]

Who hasn't happily waved at the passing engineer as a train goes by? How often are the men or women who drive the trains objectified and not seen as real people? They have the same hopes, dreams, fears and phobias as the rest of us.

Robbery

By January 1927, the electric railway system across Texas was bringing in large sums of money. W.L. Sasser, the cashier for the Galveston-Houston Electric Company's interurban ticket office in Houston, heard a noise while counting the cash after the station had closed. Upon checking the noise, Sasser came face to face with two armed robbers who then locked him in the vault and stole $2,100. Sasser was trapped for half an hour before the dispatcher heard him and freed him. The robbers fled the scene and were never found.

Stay Off the Bridge

During the Great Depression, many unemployed, destitute men drifted across the United States. In December 1932, two drifters, Earl Mueller and LeGrange England, were traveling with a man who never mentioned his name.

The trio had jumped off a train from Beaumont and were walking toward Houston, following the tracks. They began to cross a high trestle bridge when an interurban train car came barreling toward them.

Earl and LaGrange knew they could not turn around and outrun the train. Thinking quickly, they hung from the side of the bridge, gripping the bridge ties with every ounce of energy they possessed.

The man with no name was not so quick. As the rolling interurban car came closer and closer, he was either going to be shredded by the steel wheels or could jump off the bridge. The man with no name jumped.

He was found in a brushy area still alive. He was taken to the hospital but died a short time later. The only identification on his body was a letter from Seattle, Washington, with the salutation "Dear Pete," and it was signed "Harold."

May Pete rest in peace.

THE BANANA PEEL INCIDENT

BANANA PEEL
COST FT. WORTH
MAN RIGHT ARM

Headline in the *Corsicana Sun* on May 2, 1924. *Courtesy of Newspapers.com.*

The first Friday in May 1924 would never be forgotten by Bob Riley. The thirty-nine-year-old Riley had intentions of catching a ride on a Fort Worth streetcar. Unfortunately for Bob, someone had thrown away a banana peel close to the tracks.

As Bob stepped toward the car, his foot stepped on the peel, and he slipped and fell, his right arm served up on the track like a guillotine. The steel wheel of the streetcar severed Bob's right arm.[69]

Slipping on a banana peel seems like a joke, a comedic, vaudeville gag. However, it was a huge problem decades ago. When bananas first came to the United States, vendors touted the peels as "sanitary wrappers," and consumers discarded them thoughtlessly.

New York had a rash of banana peel injuries, including one death. The *New York Times*, on February 9, 1896, contained the headline, "War on the Banana Skin." At the time, Teddy Roosevelt was the president of the New York Police Department.

> *President Theodore Roosevelt of the police department summoned before him Capt. Copeland of the Delancey Street Station, Capt. Grant of the Madison Street Station, Acting Capt. Kirschner of the Elder Street Station, Acting Capt. Crilly of the Union Market Station and all the sergeants and roundsmen of those precincts. He talked to them pointedly about the prevalence of banana skins in the streets on the eastside. He explained*

the bad habits of the banana skins, dwelling particularly on its tendency to toss people into the air and bring them down with terrific force on the hard pavement. He warned the captains, sergeants and roundsmen of the necessity for keeping the streets free from banana peels, apple and potato skins and similar articles.[70]

So, beware of those banana peels, or the joke could be on you.

THE BLACK WIDOW OF FORT WORTH

On October 24, 1936, Fort Worth resident Arthur Lee Wilkens (forty-four) was crushed to death when his car was struck by a Rock Island passenger train. The train carried the mangled car one hundred feet down the track. In the first half of the twentieth century, this was, unfortunately, a common occurrence, a timeless phenomenon of technology outpacing safety. However, there is more to this story than a deadly crash.

Police officers investigating the tragedy were suspicious. What raised their suspicion is lost to history. Did evidence reveal that Wilken was in the passenger seat of the wrecked vehicle? Why was Wilkens's car sitting on the track when death arrived?

Naturally, officers questioned those close to Wilkens, including Wilkens's wife, Birdie Wilkins (forty), and Wilkens's close friend Edgar Sumrall (twenty-seven). On October 27, in the presence of District Attorney W.T. Parker and newspaper reporters, Sumrall confessed to a plot most nefarious.

Sumrall told Parker he drove Wilkens, who was asleep in the car, in front of the onrushing train. He said he expected to receive a portion of Wilkens's $1,000 life insurance.

The prisoner related how he went to Wilkens's home Saturday night and how he and Wilkens later visited several taverns and afterwards drove to Sumrall's home.

"I went in and ate supper," Sumrall told Parker. "When I returned to the car, Wilkens was asleep. I drove to the crossing and sat there waiting on a train."

When he saw the headlight of a passenger from Dallas, Sumrall said he "drove the car onto the track and ran."[71]

COUPLE CONFESSES WEIRD TEXAS SLAYING PLOT

Mrs. Birdie Wilkins, 33, and Luther Edgar Sumrall, 27 (left), told Fort Worth officers they plotted to kill her husband, Arthur Lee Wilkins, 44, whose body was found in the wreckage of an automobile (above) wrecked by a train at a grade crossing. The victim, asleep, was left in the car on the tracks, Sumrall said. He and the woman planned to divide expected compensation from the railroad. (Associated Press Photos)

Birdie Wilkins, the black widow of Fort Worth, and Luther Edgar Sumrall. Their sinister plot resulted in the wrecked automobile and death of Birdie's husband, Arthur Lee Wilkens. *Courtesy of the* Monitor *and Newspapers.com.*

Sumrall also implicated Birdie Wilkens in the crime. The next day, Birdie Wilkins admitted to her involvement as the weaver of this deadly plot. Birdie confessed that a third unnamed man had given her the idea of killing her husband a few years before. The unnamed man offered his assistance, which Birdie rejected. No one knows why Birdie decided to pursue the crime two years later. She asked Sumrall to help her carry out the deed for a portion of the life insurance money.

Later in the day, Birdie claimed that she only confessed to the crime because she had been beaten by investigators. Dr. Burke Brewster was called in by Sheriff A.B. Carter to examine Birdie. The doctor did not find any marks of violence. Investigators denied resorting to violence when questioning Birdie. Both Birdie and Sumrall were charged with murder.

On the day before Thanksgiving, Birdie Wilkens was sentenced to twenty-five years in prison. When the verdict was read, Birdie smirked at the judge, then, laughing out loud, yelled, "Ain't that grand!"[72] The outburst was most likely due to the fact that the jury had the ability to sentence her to the electric chair. Birdie later told a jailer, "Twenty-five years—that ain't so bad."[73] Birdie's lawyers immediately appealed the conviction.

In Austin, on December 8, the court of criminal appeals could not find any evidence to overturn the twenty-five-year sentence. Meanwhile, her accomplice, Edgar Sumrall, awaited his trial.

The holidays passed and the calendar turned to a new year before Sumrall's trial began. On March 9, 1937, Sumrall was convicted of murder and received a sentence of thirty years. Meanwhile, Birdie took one more stab at freedom.

After the court of criminal appeals upheld Birdie's conviction, her lawyers made a motion for a rehearing. Unfortunately for Birdie, the court of criminal appeals overruled her motion. Her sentence was finally affirmed. The black widow of Fort Worth would head to jail for the next twenty-five years.

The Mailman Has a Shotgun

Before railway transportation was introduced, mail was carried on horseback, stagecoach, sulky, canal packet and steamboat.[74] Mail was slow, postage rates were high, schedules were uncertain and, in many out-of-the-way places, deliveries were as much as three to four weeks apart.

Railroads helped build the United States and were vital to the free exchange of information and goods that became essential to the American way of life.

Railway transportation increased the frequency of deliveries, reduced postage rates and sped up the movement of mail.

The first mail transported by train was carried in 1831. In 1869, the Railway Mail Service (RMS) was inaugurated to handle the transportation and sorting of mail aboard trains called Rail Post Offices (RPO). Many American railroads earned substantial revenues through contracts with the U.S. Post Office to carry mail.

The cabinet where postal workers kept their shotgun on car 360 at the Interurban Railway Museum in Plano, Texas. *Photograph by Jeff Campbell.*

In 1914, the U.S. Post Office signed a contract with the Texas Traction Company to carry mail by rail. Cars were outfitted in the rear third to house regulation-designed post offices. The cars also carried two armed postal employees who canceled and sorted mail. The two employees dropped off and picked up mail bags at every stop.

Besides being armed with a pistol, the two employees kept a shotgun in a cabinet. Also, their work could be very hot, as the windows in the rolling post office did not open.

The pistols, shotgun and closed windows were a cumulative deterrent to bandits in a time when cash was sent through the mail before credit cards and checks.

The Murdering Train Hopper

In the winter of 1920–21, Clifford H. Rogers of Austin was train hopping freight trains from the East Coast to Los Angeles, California. In Houston, he met Harvey Hughes from Detroit, Michigan. The two men decided to travel to California together. Traveling west, the two train-hoppers were hungry and sought out a restaurant. Hughes was penniless, and Rogers bought his meal. He paid with a twenty-dollar bill and pocketed the change.

On January 24, near Alpine, a dying Rogers was found in a boxcar with a bullet wound to his back. Before he passed away, Rogers told the railroad employees that Hughes had shot him in the back and robbed him.

When Hughes was apprehended, the change from the twenty-dollar bill was found in his pocket. On February 20, 1921, Hughes was sentenced to death for the murder of Clifford H. Rogers.

Vandalism

As long as there have been railroad lines, there has been railroad vandalism. In March 1977, the House of Representatives' Subcommittee on Crime stated:

> *The Subcommittee on Crime of the House Committee on the Judiciary is beginning hearings today on what we consider to be a critical subject that has, in some ways, been ignored across the years by both branches of the Congress; the question of railroad vandalism and some of the injuries and the serious damages that are a consequence thereof. We have found out already that in 1976, there were some 42,564 acts of vandalism, which is dangerous not only to the passengers but also to people who man the trams on our Nation's railways. This vandalism cost has exceeded $5.5 million in 1976 alone.*[75]

Texas, like all states, has had its fair share of railroad vandalism. An act done in jest can unfortunately cause a lot of pain and suffering.

One of the most common acts that occurred was flipping the rail switch. The rail switch manually operated and allowed the railroad employees to redirect trains from the main line to another line. It may seem funny to send a train down the wrong track, but the results were usually devastating and disastrous.

In 1950, an open switch caused a twenty-two-car freight train to crash inside the Houston city limits. The switch had been bashed open with a thirty-inch bolt taken from a tool shed close to the accident site. Also, the batteries in the signal control box had been smashed. The train was carrying a heavy load of cement to Port Houston. As the engine careened into the air, the three crewmen aboard jumped to avoid certain death. Luckily, they survived with minor injuries.[76]

A rail switch was instrumental to the safe operation of a railway. Railway staff would flip the manual switch to redirect train cars from the main line to a siding line. This allowed other trains to pass traveling in the opposite direction. When the switch was flipped, the red flag would turn 90 degrees, making it visible to the motorman and indicating that the train was set to go to the siding station. Both mechanical and electrically operated rail switches are still used on today's railroads. *Photograph courtesy of the Johnnie J. Myers Archives.*

Fourteen years later, a similar incident occurred in Dallas. Vandals were able to hammer a lock off a switch and flip it to a spur track. Two diesel units and twelve cars were derailed, and three hundred feet of track was demolished. The estimated costs of the vandalism $100,000. Once again, there were three crewman who survived the wreckage.[77]

Placing obstructions on the railroad tracks was another dangerous act committed by vandals. Sometimes, these obstructions resulted in the desired carnage sought by the perpetrators. Other times, crisis was averted by good will and good luck.

In 1910, a man was walking by the railroad tracks two miles west of Hallsville, Texas. The tracks belonged to the Texas & Pacific Railway. The man noticed spikes driven into seven expansion joints of the steel rail. As quickly as walking could carry him, the unnamed man contacted a railroad agent.

Railroad employees were dispatched to the scene and able to intercede to prevent disaster.[78]

On the night of December 10, 1929, disaster struck. Twenty people were injured when the Crimson Limited, two interurban cars owned by the North Texas Traction Company, crashed between Dallas and Fort Worth.

J.S. Joyce, a motorman, said he felt the front truck of his car strike some obstruction, which he believed to be a piece of steel. Immediately, the first car jumped the track, plowed up the roadbed for eighty yards and then turned over. The trailer plunged on forty yards past the first car and stopped on the brink of a steep incline.

Ralph N. Smith of Muskogee, Oklahoma, a passenger in the rear car, was among those who gave eyewitness accounts of the wrecks. "I had just remarked that there would be nothing but a grease spot left if our train jumped the track at the speed it was going….Just then, I felt a severe jolt, and as the cars left the track, I heard a woman scream, 'Save my baby.' When the cars finally stopped, I could see she still held her child and that, apparently, the latter was not seriously hurt."

Edmand Silverbrand of Detroit, another passenger, said he had just finished traveling ten thousand miles without mishap in an airplane. "I feel I'm the luckiest man in Dallas….I don't see how anyone escaped alive. The whole inside of the car seemed to be torn to pieces, and broken glass was scattered everywhere. I helped to get out one baby whose mother seemed to be injured."[79]

Short-circuited electric wires popped, flashed and slithered like snakes while the sirens of twelve ambulances and three firetrucks added to the chaotic symphony surrounding the injured and the first responders.

The accident was investigated by T.H. Wren, a claim agent for the Northern Texas Traction Company, and Superintendent Horace Floyd of the Texas Electric Railway. The investigation revealed that an express car had safely passed the same section of track a short time before the wreck. Between the time of the express car passing and the wreck, children playing with rocks were seen near the track.

After the wreck, broken rocks were found scattered along the rail among the wrecked cars and telephone poles that were snapped like matchsticks. A childish prank had resulted in massive carnage. Twenty people were injured, but luckily, no one died.

However, on August 31, 1930, death would not take a holiday. The Texas Special was one of the most famous passenger trains of its time. The Texas Special was jointly operated by the Frisco (the St. Louis–San Francisco

Railway) and the Katy (the Missouri-Kansas Railway). The Texas Special's route operated between San Antonio and St. Louis.

On the night of August 31, the Texas Special crashed just outside of St. Louis, Missouri. Six people were killed, and fifty-eight others were injured. Railroad damages were estimated to reach $1 million.

What caused the wreck of the Texas Special? Stones placed on the tracks along a long curve. The culprits, never found, were believed to be potential train robbers who likely only wanted to stop the train or someone with a grudge against the railroad.

22
EVER-PRESENT DANGER

Most of the stories in this book come from a time that seems far away. It would be nice to think that railroad accidents are a thing of the past. However, danger and mayhem are always just around the bend. Railroad deaths continued into the new millennium.

- 2004, San Antonio, Texas: Lethal chlorine gas leaked from a railroad car when a Union Pacific train collided with a Burlington Northern-Santa Fe train. Fifty people were injured and four died, mainly from the gas leak.
- 2006, Austin, Texas: Miss Deaf America, Tara Rose McAvoy (eighteen) was walking parallel with the railroad tracks, texting her parents. The oncoming train sounded its horn and tried to stop. McAvoy was struck and killed by the train's snowplow, which extended sixteen inches past each side of the tracks.
- 2012, Midland, Texas: A parade float carrying veterans to a banquet in their honor became trapped behind another car. Sitting helplessly on tracks, the float was hit by an oncoming Union Pacific train. Sixteen veterans were injured and four died.
- 2015, Odessa, Texas: A prison bus trying to navigate winter conditions slid off an icy road and into a moving train. Two officers and eight inmates were killed.

NOTES

About the Interurban Railways

1. William D. Middleton, "Goodbye to the Interurban," American Heritage, https://www.americanheritage.com/goodbye-interurban.

Chapter 2

2. "Ex-Waxa Man Slain in Dallas and Laid to Rest Here Monday," *Waxahachie Daily Light*, November 5, 1928, www.newspapers.com.
3. "Dewey Hunt's Five-Year Fight Is Ended with Jaunty Walk to Chair," *Corsicana Daily Sun*, December 29, 1933.
4. "Mountaineer Goes to Chair," *Knoxville Journal*, December 30, 1933.

Chapter 3

5. Ibid.
6. Allan Turner, "Eternity's Gate Slowly Closing at Peckerwood Hill," Chron., August 3, 2021, https://www.chron.com/news/houston-texas/article/Eternity-s-gate-slowly-closing-at-Peckerwood-Hill-3761731.php.
7. Walls Unit was the nickname for the Huntsville Unit of the Texas Department of Criminal Justice; Robyn Ross, "Laid to Rest in Huntsville,"

Texas Observer, March 11, 2014, https://www.texasobserver.org/prison-inmates-laid-rest-huntsville/.

8. Peter Applebome, "Texas Town Leading in Executions in a New U.S. Era of Death Penalty," *New York Times*, September 6, 1986, https://www.nytimes.com/1986/09/06/us/texas-town-leading-in-executions-in-a-new-us-era-of-death-penalty.html.
9. Ross, "Laid to Rest."
10. Peter Applebome, "Texas Prisoner Burials Are a Gentle Touch in a Punitive System," *New York Times*, January 5, 2012, https://www.nytimes.com/2012/01/05/us/texas-prisoner-burials-are-a-gentle-touch-in-a-punitive-system.html.

Chapter 4

11. "Woman, Shot by Former Husband, Better; Bail Made," *Fort Worth Star Telegram*, February 24, 1921, www.newspapers.com.
12. Ibid.
13. "No Bill Is Voted for Noel Killing," *Fort Worth Star Telegram*, March 19, 1921, www.nespapers.com.

Chapter 6

14. "Railway Pamphlets," vol. 109 (1878), 109.
15. Clinton Bradford Herrick, MD, *Railway Surgery: A Handbook on the Management of Injuries* (New York: William Wood and Company, 1899), 11–12.
16. Ibid., 3.

Chapter 8

17. Email interview with Interurban Railway Museum archive manager Debbie Calvin, August 6, 2021.
18. "Interurban Collided with Automobile," *Wichita Daily Times*, October 11, 1912, www.newspapers.com.
19. Interurban cars were similar to trolleys.

Chapter 11

20. "Space Purchased by Mr. and Mrs. A.P. Waters to Thank Friends and Family," *Fort Worth Record*, October 23, 1920.
21. Deah Berry Mitchell, "History of Como Fort Worth," Visit Fort Worth, https://www.fortworth.com/blog/post/history-of-como-fort-worth/.
22. "Barnes Waves Hearing: Bond Asked, Denied," *Fort Worth Star Telegram*, October 30, 1920.
23. "Barnes Case Goes to Jury; Death Penalty Is Asked," *Fort Worth Star Telegram*, November 17, 1920.
24. Ibid.
25. "Man Held Now for Huntsville Prison," *Galveston Daily News*, October 18, 1926.
26. Texas State Library and Archives Commission, "Pardons and Paroles," https://www.tsl.texas.gov/exhibits/prisons/inquiry/pardons.html.
27. Texas State Library, "Do Such Acts of Fergusonism," https://www.tsl.texas.gov/sites/default/files/public/tslac/exec/documents/struggles6_2_22_701_fergusonism.pdf.

Chapter 12

28. El Paso Streetcar, "Moving Forward," https://www.crrma.org/streetcar.
29. Ibid.
30. "New Brewery Will Soon be Making Beer," *El Paso Herald*, June 24, 1921.
31. Fred Minnick, "Cross Border Bourbon," *Whiskey Magazine*, n.d., https://whiskymag.com/story?cross-border-bourbon.
32. "A New Way to Get a Drink in El Paso," *El Paso Herald*, May 20, 1922.
33. "Three-Year-Old Runs Streetcar off Tracks," *El Paso Herald*, May 4, 1922.
34. "Youthful Imposter Works with Boy Robbers, Say El Paso Policemen," *El Paso Herald*, March 23, 1923.
35. "Trolley Bandit Robs Motorman in Bold Holdup," *El Paso Herald*, January 18, 1920.
36. "Trolley Bandit Idle for Month Stages Robbery," *El Paso Herald*, March 17, 1920.
37. "Plea for Family Hits Highwayman: Gives Back Money," *El Paso Herald*, October 3, 1921.
38. Ibid.

39. "Trolley Death Damage Suit Trial Postponed," *El Paso Herald*, November 17, 1953.
40. anylaw, "*James Lamar Webb v. State*," https://www.anylaw.com/case/james-lamar-webb-v-state/court-of-criminal-appeals-of-texas/05-19-1954/U8kfYGYBTlTomsSBmSbT.
41. Ibid.

Chapter 13

42. John Minton, "'The Waterman Train Wreck': Tracking a Folksong in Deep East Texas," *Journal of Folklore Research* 28, no. 2–3 (1991): 179–219. http://www.jstor.org/stable/3814503.

Chapter 14

43. "Wreck of the Old 97," Public domain. The song is about the 1903 train wreck near Danville, Virginia. The song was first recorded by G.B Grayson and Henry Whitter. Vernon Dalhart, from Jefferson, Texas, recorded the song in 1924. Dalhart's version is widely regarded as the first million-selling country recording. The origin of the song is open to debate. The music is taken from a song written in 1865 by Henry Clay Work, "The Ship That Never Returned." Lyrics are claimed by multiple individuals.
44. Jeff Campbell, "Jefferson Train Wrecks," Stephen F. Austin State University, August 2013, https://www.sfasu.edu/heritagecenter/7899.asp.
45. *Marshall News Messenger*, August 12, 1947.

Chapter 15

46. Benjamin Heber Johnson, *Revolution in Texas: How a Forgotten Rebellion and Its Bloody Suppression Turned Mexicans into Americans* (New Haven, CT: Yale University Press, 2005), 32.
47. Smoker cars were passenger cars where passengers could smoke pipes, cigars or cigarettes.
48. *Austin American Sun*, October 20, 1915.

49. Rebecca Onion, "America's Lost History of Border Violence," May 5, 2016, https://slate.com/news-and-politics/2016/05/texas-finally-begins-to-grapple-with-its-ugly-history-of-border-violence-against-mexican-americans.html.
50. "During Three Months 21 Americans Killed in Border Trouble," *Austin American Sun*, October 31, 1915.
51. "Mexican Bandits Fire on Patrol Near Brownsville," *Houston Post*, March 16, 1916.
52. "Three Mexican Bandits Killed and Three Captured by Posse," *Laredo Weekly Times*, June 18, 1916.
53. Stockmen were ranch owners or ranch foremen.
54. National Archives, "The Zimmerman Telegram," https://www.archives.gov/education/lessons/zimmermann.

Chapter 16

55. "Bodies Strewn Along Track for 500 Feet by Impact of Blow," *Brownsville Herald*, March 14, 1940.
56. "27 Killed in Valley Crash," *Valley Morning Star*, March 15, 1940.

Chapter 17

57. Larry Tye, *Rising from the Rails: Pullman Porters and the Making of the Black Middle Class* (New York: Holt & Company Press, 2004), xi.
58. *Dom Flemons Presents Black Cowboys* (Washington, D.C.: Smithsonian Folkways Recordings, 2018), liner notes.
59. *Fort Worth Star Telegram*, September 29, 1917.

Chapter 18

60. *El Paso Herald*, June 11, 1907.
61. *Austin American Statesman*, June 11, 1907.

Chapter 19

62. *Courier Gazette*, March 31, 1921.

Chapter 20

63. Johnnie J. Myers, *Central Electric Railfans' Association Texas Electric Railway*, bulletin 121, edited by LeRoy O. King Jr. (Chicago: Central Electric Railfans Association, n.d.). The Central Electric Railfans Association (CERA) was formed 1938. The CERA encourages the study of the history, equipment and operations of urban, suburban, interurban and main line electric railways.
64. Ibid.
65. "Interurban Crash Near Dallas Hurts 50," *Del Rio News Herald*, April 11, 1948.
66. "50 Persons Injured as Two Interurbans Crash," *Brownsville Herald*, April 11, 1948.

Chapter 21

67. "Railroad Man Kills Himself," *Houston Post*, February 29, 1908.
68. "Motorman Is Suicide," *Valley Morning Star*, September 1, 1940.
69. "2 Corsicana Teachers Killed," *Corsicana Daily Sun*, May 2, 1924, www.newspapers.com.
70. "War on the Banana Skin," *New York Times*, February 9, 1896, https://timesmachine.nytimes.com/timesmachine/1896/02/09/108222329.pdf?pdf_redirect=true&ip=0.
71. "Charge Filed Against Two for Killing" *Valley Morning Star*, October 27, 1936.
72. "Birdie Wilkins Is Guilty, Sentenced to 25-Year Term," Marshall News Messenger, November 26, 1936.
73. Ibid.
74. A sulky is a lightweight cart, pulled by a horse.
75. Office of Justice Programs, "Railroad Vandalism," March 9, 1977, https://www.ojp.gov/pdffiles1/Digitization/77472NCJRS.pdf.
76. "Vandals Cause Rail Wreck," *Baytown*, July 11, 1950.
77. "Blame Vandals in Rail Wreck," *Corsicana Daily*, July 20, 1964.
78. "Vandals Attempt to Derail Train Rails are Spiked," *Fort Worth Record and Register*, December 6, 1910.
79. "Rocks Are Blamed for Interurban Wreck," *Waxahachie Daily Light*, December 11, 1929.

The Traveling Motormen at the Interurban Railway Museum. The Traveling Motormen teach children, civic groups and seniors about the history of Texas rail. The Interurban Railway Museum is located at 901 East Fifteenth Street in historic downtown Plano, Texas. (*Left to right*): Terry Fleming, Harold Larson and Jeff Campbell. *Photograph by Charlotte Loontiens.*

Special thanks to Austin Ng (*left*) and Hunter Herring (*right*) for their contributions to this book. *Courtesy of Charlotte Loontiens.*

ABOUT THE AUTHOR

Author Jeff Campbell at the 2019 Preservation Texas Awards in Austin's Paramount Theatre. *Photograph courtesy of the Plano Conservancy for Historic Preservation.*

Jeff Campbell is the executive director of the Plano Conservancy for Historic Preservation. He writes about Plano history for *Plano Magazine* and also coauthored *Football and Integration in Plano, Texas: Stay in There, Wildcats!* (The History Press, 2014), *Plano's Historic Cemeteries* (Arcadia, 2014), *Hidden History of Plano* (The History Press, 2020) and *Texas Bluegrass History* (The History Press, 2021). Jeff has worked on historic preservation projects in Texas, Louisiana and New Mexico. He serves on the board of the Texas chapter of the Association of Gravestone Studies and on the advisory board of Texas Dance Hall Preservation, and he is a member of the Forest Fire Lookout Association. In 2019, he received a Preservation Texas Award for his work at Historic Davis Cemetery.

Since 2001, the Plano Conservancy for Historic Preservation has managed the Interurban Railway Museum in Historic Downtown Plano, Texas.

The Plano Conservancy also works to preserve Plano's historic cemeteries and produce exhibits, books, wayside signs and events, such as the Hike Through History and an annual Archaeology Fair.

ABOUT THE PLANO CONSERVANCY FOR HISTORIC PRESERVATION INC.

Photograph courtesy of the Plano Conservancy for Historic Preservation.

The Plano Conservancy for Historic Preservation Inc., along with the City of Plano, maintains the Interurban Railway Museum and has implemented projects to improve the exhibits at the museum. The City of Plano owns the museum. The Plano Conservancy for Historic Preservation Inc. staffs, operates and houses its offices there and has been instrumental in developing and maintaining the museum and its exhibits. Also a part of the Interurban Railway Museum is the Johnnie J. Myers Research Center, which contains photographs, books, oral histories, visual media, documents and artifacts.